HEINEMANN MODULAR MATHEMATICS
for
EDEXCEL AS AND A-LEVEL

Mechanics 2

John Hebborn   Jean Littlewood

1
2
3
4
5

Heinemann Educational Publishers,
a division of Heinemann Publishers (Oxford) Ltd,
Halley Court, Jordan Hill, Oxford, OX2 8EJ

OXFORD MELBOURNE JOHANNESBURG AUCKLAND
BLANTYRE IBADAN GABORONE PORTSMOUTH NH (USA)
CHICAGO

First published 2000

01 02 10 9 8 7 6 5 4

ISBN 0 435 51075 4

Cover design by Gecko Limited

Original design by Geoffrey Wadsley: additional design work by Jim Turner

Typeset and illustrated by Tech-Set Ltd, Gateshead, Tyne and Wear

Printed in Great Britain by The Bath Press, Bath

Acknowledgements:

The publisher's and authors' thanks are due to the Edexcel for permission to reproduce questions from past examination papers. These are marked with an [E].

The answers have been provided by the authors and are not the responsibility of the examining board.

About this book

This book is designed to provide you with the best preparation possible for your Edexcel M2 exam. The series authors are senior examiners and exam moderators themselves and have a good understanding of Edexcel's requirements.

Finding your way around

To help you find your way around when you are studying and revising use the:

- **edge marks** (shown on the front page) – these help you to get to the right chapter quickly;
- **contents list** – this lists the headings that identify key syllabus ideas covered in the book so you can turn straight to them;
- **index** – if you need to find a topic the **bold** number shows where to find the main entry on a topic.

Remembering key ideas

We have provided clear explanations of the key ideas and techniques you need throughout the book. Key ideas you need to remember are listed in a **summary of key points** at the end of each chapter and marked like this in the chapters:

■ $$e = \frac{v}{u} \text{ or } v = eu$$

Exercises and exam questions

In this book questions are carefully graded so they increase in difficulty and gradually bring you up to exam standard.

- **past exam questions** are marked with an E;
- **review exercises** on pages 83 and 145 help you practise answering questions from several areas of mathematics at once, as in the real exam;
- **exam style practice paper** – this is designed to help you prepare for the exam itself;
- **answers** are included at the end of the book – use them to check your work.

Contents

3 Work, energy and power

4 Collisions

5 Statics of rigid bodies

Kinematics of a particle moving in a straight line or plane

1

In Book M1 chapter 3 you learnt about the motion of a particle moving in a straight line. Section 3.2 of that chapter dealt with the motion of a particle moving in a vertical line under the action of gravity. However, it is relatively unusual for a particle to be moving only in a vertical direction; most particles are thrown at an angle, which may be very small, to the vertical. In the first section of this chapter you will learn more about the motion of a particle through the air.

1.1 Projectiles

Ignoring air resistance, a particle thrown into the air moves freely under gravity. Such a particle is generally called a **projectile**. It moves through the air along a curved path.

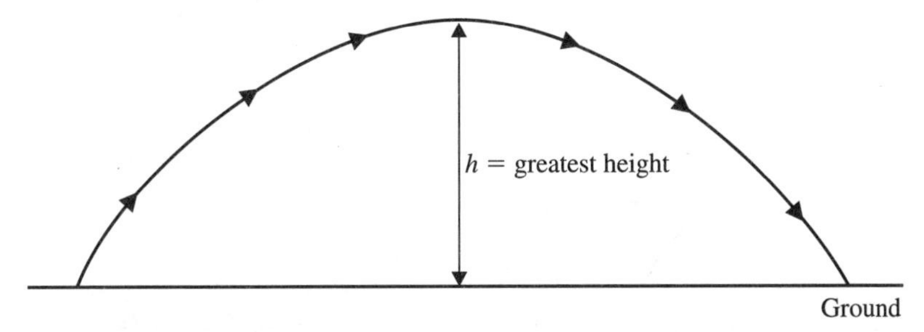

A projectile has an acceleration of $g = 9.8\,\text{m}\,\text{s}^{-2}$ vertically downwards. Problems involving projectiles can be solved by considering the vertical and horizontal motions separately. As the acceleration is entirely vertical the projectile has no acceleration horizontally and so its horizontal velocity is unchanged throughout the motion. Vertically, the motion can be investigated by applying the uniform acceleration equations as in Book M1 section 3.2.

Horizontal projection

A stone which is thrown horizontally from the top of a cliff will fall into the sea below, but will travel horizontally as well as vertically while falling.

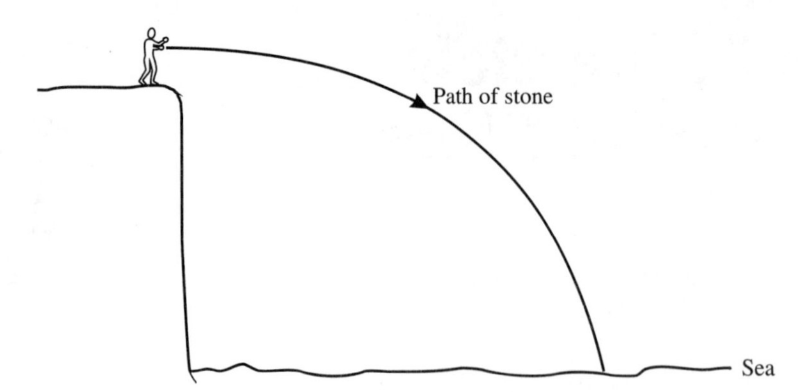

The stone begins its journey at the highest point of the projectile path shown in the previous diagram. At this highest point the stone's vertical velocity is zero but it has a vertical acceleration, g.

Example 1

A particle is projected horizontally with speed $15\,\text{m}\,\text{s}^{-1}$ from a point 24 m above a horizontal surface. Calculate to three significant figures the time taken for the particle to reach the surface and the horizontal distance travelled in that time.

Since the vertical distance to the plane is known, use the vertical motion to find the time taken to reach the surface. As in Book M1 section 3.2 choose a direction to be positive.

The vertical motion is downwards, so take this direction to be positive.

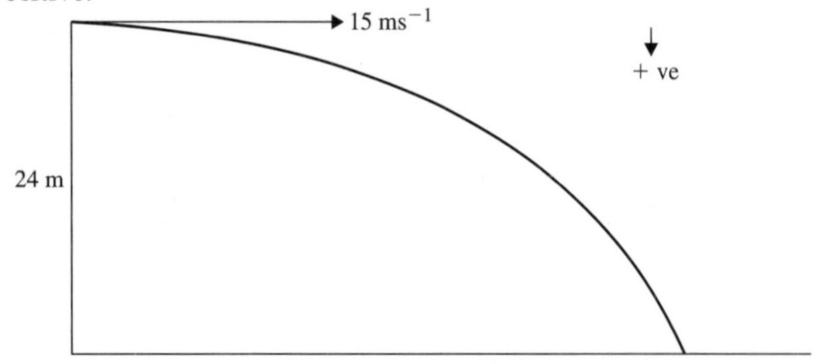

As the particle starts with a horizontal velocity its initial vertical speed is zero.

Known quantities are: $a = 9.8\,\text{m}\,\text{s}^{-2}$

$\downarrow$ $s = 24$ m

+ve $u = 0\,\text{m}\,\text{s}^{-1}$

Using: $s = ut + \frac{1}{2}at^2$

Gives:

$$24 = 0 \times t + \tfrac{1}{2} \times 9.8 \times t^2$$

$$t^2 = \frac{48}{9.8}$$

$$t = 2.213$$

The time taken to reach the surface is 2.21 s.

Horizontal motion must be considered to find the horizontal distance travelled. As there is no horizontal acceleration the horizontal velocity is unchanged.

Hence: distance travelled = (horizontal speed) × time

$$= (15 \times 2.213)$$
$$= 33.19$$

The horizontal distance travelled is 33.2 m.

Example 2

A particle is projected horizontally with a speed of $19.6\,\text{m s}^{-1}$. Find the distance of the particle from its starting point after 1.5 seconds.

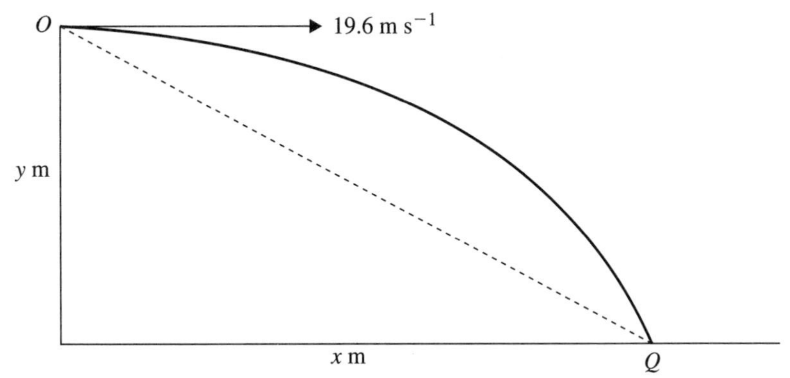

The distance between the particle's position Q when $t = 1.5$ s and its starting point O is shown by the dotted line in the above diagram. You need to find the horizontal and vertical distances, x m and y m.

Consider the horizontal motion.

As there is no horizontal acceleration:

$$x = (\text{horizontal speed}) \times \text{time}$$
$$= (19.6 \times 1.5) = 29.4$$

The distance x is 29.4 m.

Now consider the vertical motion.

Known quantities are: $u = 0\ \text{ms}^{-1}$

$\downarrow$ $a = 9.8\,\text{m s}^{-2}$

+ve $t = 1.5\,\text{s}$

Using: $s = ut + \frac{1}{2}at^2$

Gives: $y = 0 \times 1.5 + \frac{1}{2} \times 9.8 \times 1.5^2$

$y = 11.025$

The distance y is 11.025 m.

By Pythagoras' Theorem:

$$OQ = \sqrt{(x^2 + y^2)} = \sqrt{(29.4^2 + 11.025^2)}$$
$$= 31.39$$

The particle is 31.4 m from its point of projection after 1.5 seconds.

Example 3

A stone is thrown horizontally with a speed of $20\,\text{m}\,\text{s}^{-1}$. By modelling the stone as a particle find the magnitude and direction of the stone's velocity 2 seconds later.

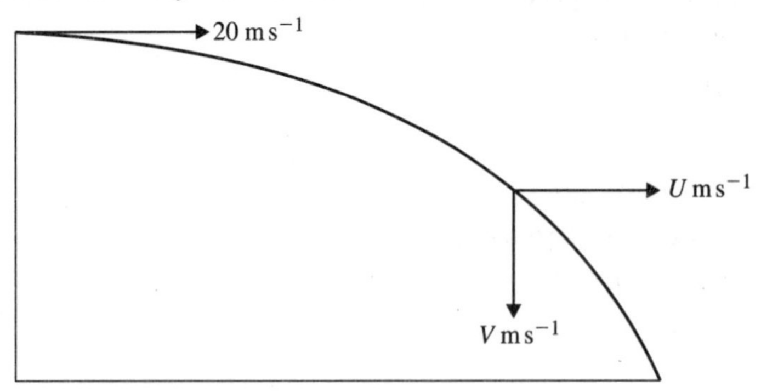

To find the magnitude and direction of the stone's velocity after 2 seconds you need the resultant of the horizontal and vertical components of the velocity.

Let the components be $U\,\text{m}\,\text{s}^{-1}$ and $V\,\text{m}\,\text{s}^{-1}$ as shown in the diagram.

For horizontal motion there is no acceleration so $U = 20$.

For vertical motion the known quantities are:

$$\downarrow \quad t = 2\,\text{s}$$
$$+\text{ve} \quad u = 0\,\text{m}\,\text{s}^{-1}$$
$$a = 9.8\,\text{m}\,\text{s}^{-2}$$
$$v = V\,\text{m}\,\text{s}^{-1}$$

Using: $\quad v = u + at$

Gives: $\quad V = 0 + 9.8 \times 2 = 19.6$

The two components of the stone's velocity are perpendicular. Their resultant is the diagonal of the rectangle shown opposite.

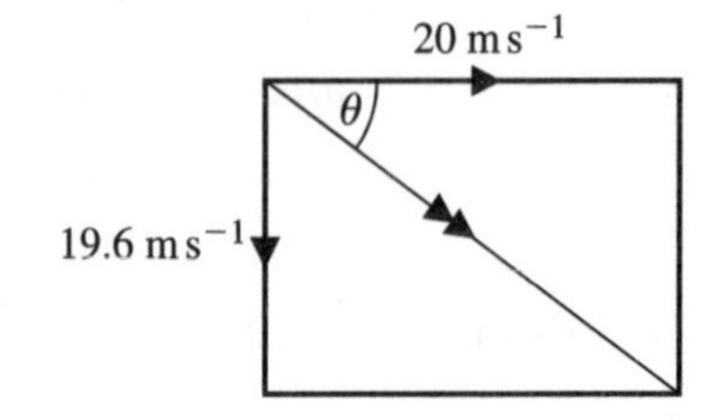

The magnitude of the resultant $= \sqrt{(20^2 + 19.6^2)} = 28.0$.

As: $\quad \tan\theta = \dfrac{19.6}{20}$

Hence: $\quad \theta = 44.4^\circ$

The stone's velocity has magnitude $28\,\text{m}\,\text{s}^{-1}$ and its direction is 44.4° below the horizontal.

Projectiles using vector notation

Information about a projectile can be conveyed in vector form with **i** and **j** being the unit vectors in the horizontal and upward vertical directions respectively. Separate the horizontal and vertical information and solve as before.

Example 4

A particle is projected at $t = 0$ with a velocity $5\mathbf{i}\,\text{m s}^{-1}$ from a point with position vector $(6\mathbf{i} + 50\mathbf{j})$ m relative to a fixed origin O where $\mathbf{i}$ and $\mathbf{j}$ are unit vectors in the horizontal and upward vertical directions respectively. Find the position vector of the particle 3 seconds later.

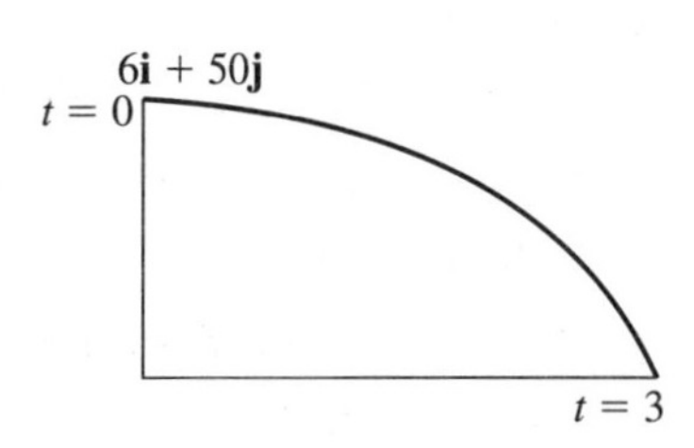

There is no horizontal acceleration.

So: Distance = speed × time

$$s = (5 \times 3) = 15$$

The horizontal distance travelled in vector form is $15\mathbf{i}$.

For vertical motion known quantities are:

$$\downarrow \quad a = 9.8\,\text{m s}^{-2}$$

$$u = 0\ \text{ms}^{-1}$$

Using: $s = ut + \frac{1}{2}at^2$

Gives: $s = \frac{1}{2} \times 9.8 \times 3^2 = 44.1$

The vertical distance is downwards. In vector form it is therefore $-44.1\mathbf{j}$.

The total displacement is $15\mathbf{i} - 44.1\mathbf{j}$.

Using the vector diagram opposite gives:

$$\text{final position vector} = (6\mathbf{i} + 50\mathbf{j}) + (15\mathbf{i} - 44.1\mathbf{j})$$
$$= 21\mathbf{i} + 5.9\mathbf{j}$$

The position vector after 3 seconds is $(21\mathbf{i} + 5.9\mathbf{j})$ m.

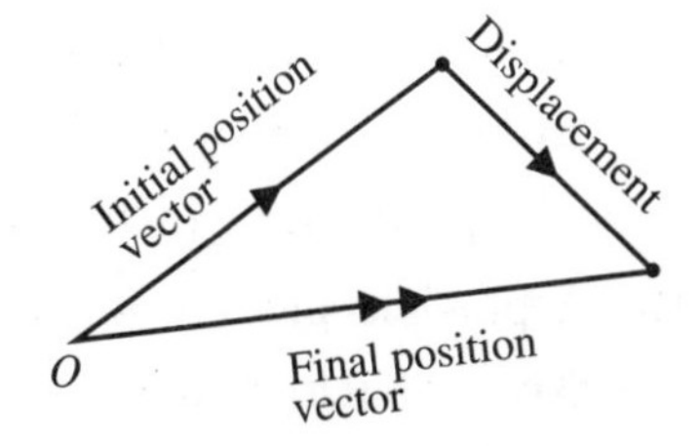

Exercise 1A

Whenever a numerical value of g is required take $g = 9.8\,\text{m s}^{-2}$.

1 A particle is projected horizontally at $15\,\text{m s}^{-1}$ from a point 60 m above a horizontal surface. Find how long it takes before the particle strikes the surface and the horizontal distance travelled in that time.

2 A particle is projected horizontally at $20\,\text{m s}^{-1}$. It strikes a horizontal surface after travelling 50 m horizontally. Find the height of the point of projection above the surface.

3 A stone is thrown horizontally from a window 3.6 m above horizontal ground. It hits the ground after travelling 10 m horizontally. Find the speed of projection.

4 A stone is thrown horizontally at $25\,\text{m}\,\text{s}^{-1}$ from the edge of a vertical cliff which is 50 m high. By modelling the stone as a particle calculate its distance from the foot of the cliff when it enters the water.

5 A bird flying horizontally at $30\,\text{m}\,\text{s}^{-1}$ at a height of 35 m drops a stone. Model the stone as a particle and hence determine how long it takes the stone to reach the ground. What is the horizontal distance travelled by the stone during this time?

6 A stone is thrown horizontally and hits the ground 7 m from the thrower 0.5 s later. Find the speed and height of projection.

7 At time $t = 0$ a particle is projected with a velocity $4\mathbf{i}\,\text{m}\,\text{s}^{-1}$ from a point with position vector $24\mathbf{j}\,\text{m}$ where $\mathbf{i}$ and $\mathbf{j}$ are unit vectors in the horizontal and upward vertical directions respectively. Find the position vector of the particle after 1.5 s.

8 At time $t = 0$ a particle is projected horizontally from a point with position vector $(x\mathbf{i} + y\mathbf{j})\,\text{m}$. Two seconds later it passes through the point with position vector $(8\mathbf{i} + 2\mathbf{j})\,\text{m}$. If the speed of projection is $4\,\text{m}\,\text{s}^{-1}$ find the values of x and y.

9 A child throws a ball horizontally at a wall 4 m away. The ball strikes the wall 0.4 m below the level of projection. Calculate the speed with which it was projected.

10 A particle slides along a horizontal table and over the edge. Given that the table is 1 m high and the speed of the particle on the table is $14\,\text{m}\,\text{s}^{-1}$ calculate (a) the vertical velocity with which the particle hits the floor (assumed horizontal) (b) the speed and direction of motion of the particle when it hits the floor.

11 A boy standing on a garage roof throws a ball horizontally with a speed of $15\,\text{m}\,\text{s}^{-1}$. The ball just clears a wall 1.5 m high 10 m away. Model the ball as a particle and hence calculate the height of projection.

12 A particle is projected horizontally at $25\,\text{m}\,\text{s}^{-1}$. Find the horizontal and vertical components of the particle's velocity 1.5s later. Find the speed of the particle and its direction of motion at that time.

13 At time $t = 0$ a particle is projected with a velocity $4\mathbf{i}\,\text{m}\,\text{s}^{-1}$ from a point with position vector $(6\mathbf{i} + 2\mathbf{j})\,\text{m}$. Find its position vector after 2 seconds.

14 At time $t = 0$ a particle is projected horizontally from a point A with position vector $(5\mathbf{i} + 30\mathbf{j})$ m. It passes through point B with position vector $(17\mathbf{i} + 10.4\mathbf{j})$ m. Find its speed of projection and the time taken to travel from A to B.

15 An aeroplane is flying horizontally at $400\,\text{m s}^{-1}$. A package is released and travels a distance of 2000 m horizontally before hitting the ground. Model the package as a particle and hence find the height of the aeroplane above the ground.

16 A batsman hits a ball horizontally with a speed of $21\,\text{m s}^{-1}$ at 1 m above the ground. Find the distance travelled horizontally by the ball before it reaches the ground.

17 A child throws a ball horizontally from a window 5 m above horizontal ground. The ball just clears a vertical wall 2.5 m high and 12 m from the house. By modelling the ball as a particle calculate the speed of projection. The ball hits the ground at Q. Calculate the distance of Q from the wall.

18 A tennis ball is served horizontally at a speed of $24\,\text{m s}^{-1}$ from a height of 2.7 m. The net is 1 m high and 12 m horizontally from the server. Model the ball as a particle and hence determine whether the ball clears the net and if so by what distance.

Projection at an angle to the horizontal

It is more usual for particles to be projected at an angle to the horizontal rather than horizontally.

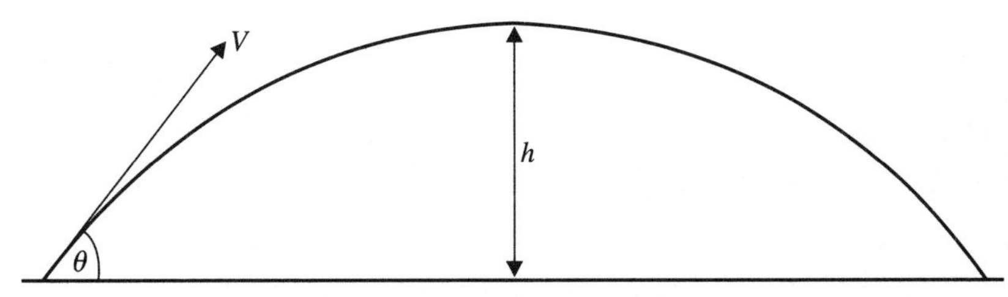

It is easier to investigate the motion of such a particle if you deal with its horizontal and vertical movement separately.

If a particle is projected with initial speed V at an angle θ to the horizontal the horizontal component of the initial velocity can be expressed as $V \cos \theta$. This remains constant throughout the motion. The vertical component of the initial velocity is $V \sin \theta$ which does not remain constant because of the vertical acceleration due to gravity.

- **For initial velocity:** **horizontal component** $= V \cos \theta$
  and: **vertical component** $= V \sin \theta$.

Example 5

A particle is projected from a point O with a speed of $30\,\text{m}\,\text{s}^{-1}$ at an angle of elevation of $\arcsin \frac{3}{5}$ (in other words the angle of elevation is θ where $\sin\theta = \frac{3}{5}$).

(a) Find the greatest height above O reached by the particle
(b) The particle strikes the horizontal through O at Q. Find the distance OQ.

(a) The vertical speed when the particle reaches its greatest height is zero.
Its initial speed is:

$$(30\sin\theta)\,\text{m}\,\text{s}^{-1} = (30 \times \tfrac{3}{5})\,\text{m}\,\text{s}^{-1} = 18\,\text{m}\,\text{s}^{-1}$$

Known quantities for vertical motion are:

$$+\text{ve}\ \uparrow \quad u = 18\,\text{m}\,\text{s}^{-1}, \quad v = 0\,\text{m}\,\text{s}^{-1}, \quad a = -9.8\,\text{m}\,\text{s}^{-2}$$

Let the greatest height be h m.

Using: $v^2 = u^2 + 2as$

Gives: $0 = 18^2 + 2 \times (-9.8)\,h$

$$h = \frac{18^2}{2 \times 9.8} = 16.53$$

The greatest height is 16.5 m.

(b) To calculate the horizontal distance you need both the velocity and the time taken. When the particle reaches Q, its vertical displacement from its starting point is zero. Use this to find the length of time the particle is in the air – often called **the time of flight**.

To find this time of flight, consider the vertical motion.

Known quantities for vertical motion are:

$$+\text{ve}\ \uparrow \quad u = 18\,\text{m}\,\text{s}^{-1}, \quad s = 0\,\text{m}, \quad a = -9.8\,\text{m}\,\text{s}^{-2}$$

Using: $s = ut + \frac{1}{2}at^2$

Gives: $0 = 18t - \frac{1}{2} \times 9.8t^2$

Factorising gives: $0 = t(18 - 4.9t)$
So either: $t = 0$ (this is the moment of projection)

or: $t = \dfrac{18}{4.9} = 3.673$ (this is the time when the particle reaches Q.)

It takes 3.673 s to travel from O to Q.

Since there is no horizontal acceleration:

$$\text{horizontal distance} = \text{speed} \times \text{time}$$

The horizontal velocity is $30 \cos \theta \,\text{m s}^{-1}$.

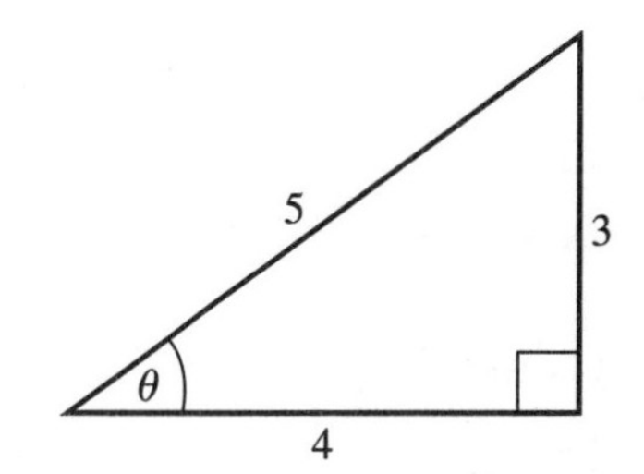

As $\sin \theta = \frac{3}{5}$, a (3, 4, 5) triangle shows that $\cos \theta = \frac{4}{5}$.

So:
$$\begin{aligned} OQ &= (30 \cos \theta \times 3.673) \\ &= (30 \times \tfrac{4}{5} \times 3.673) \\ &= 88.16 \end{aligned}$$

The distance travelled horizontally is 88.2 m. This distance is horizontal often called the **range of the projectile**.

Example 6

A girl hits a ball at an angle of $\arctan \frac{3}{4}$ to the horizontal (that is at an angle θ where $\tan \theta = \frac{3}{4}$) from a point O which is 0.5 m above level ground. The initial speed of the ball is $15 \,\text{m s}^{-1}$. The ball just clears a fence which is a horizontal distance of 18 m from the girl, as shown in the diagram. By modelling the ball as a projectile find the time taken for the ball to reach the fence and the height of the fence.

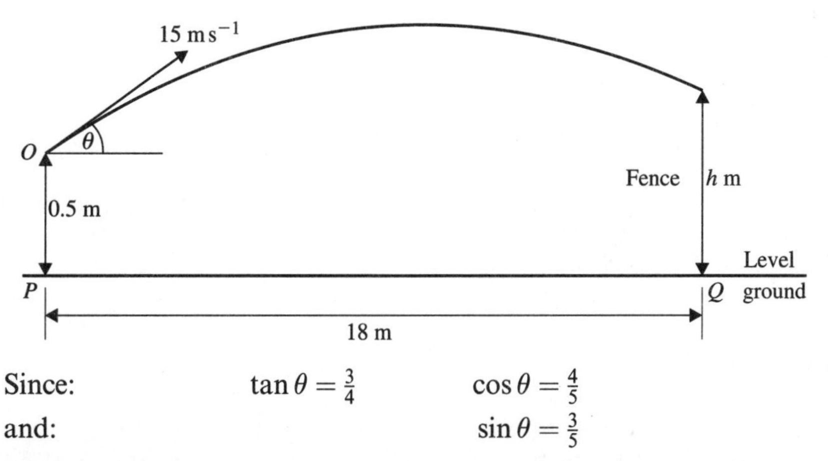

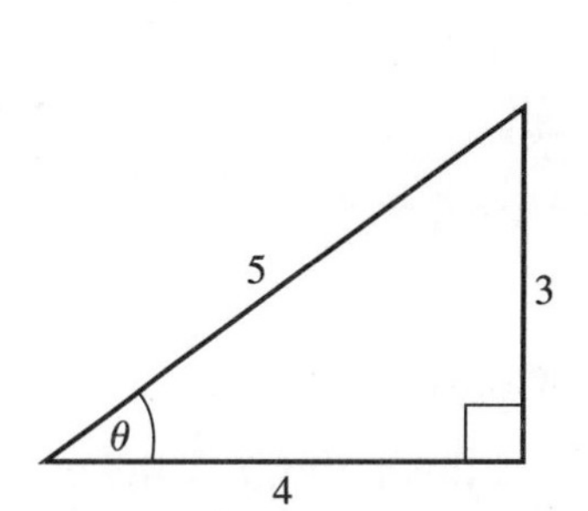

Since: $\tan \theta = \frac{3}{4}$ $\qquad \cos \theta = \frac{4}{5}$

and: $\qquad \sin \theta = \frac{3}{5}$

In this example, the horizontal speed and distance PQ are known, so the time to reach the top of the fence can be found.

The horizontal speed is $(15 \cos \theta) = (15 \times \frac{4}{5}) = 12 \,\text{m s}^{-1}$.

Since horizontal distance = 18 m:

$$\text{time} = \frac{\text{distance}}{\text{speed}} = \frac{18}{12} = 1.5$$

The ball takes 1.5 s to reach the fence.

Let the fence be h metres high. At the moment when the ball just clears the fence it is $(h - 0.5)$ m higher than its starting point.

Known quantities for vertical motion are:

+ve ↑

$s = (h - 0.5)\text{ m}$

$u = 15\sin\theta\text{ m s}^{-1} = 15 \times \frac{3}{5}\text{ m s}^{-1} = 9\text{ m s}^{-1}$

$t = 1.5\text{ s}$

$a = -9.8\text{ m s}^{-2}$

Using: $s = ut + \frac{1}{2}at^2$

Gives:

$$h - 0.5 = 9 \times 1.5 + \tfrac{1}{2}(-9.8) \times 1.5^2$$

$$h = 0.5 + 9 \times 1.5 - 4.9 \times 1.5^2$$

$$h = 2.95$$

The fence is 2.95 m high.

Example 7

A ball is projected from O at the top of a cliff with speed $V\text{ m s}^{-1}$ at an angle of 30° above the horizontal. The greatest height reached above the horizontal plane through O is 4.9 m. Show that $V = 19.6$. Given that O is 60 m above the surface of the sea, find the time taken for the ball to travel from O to the sea's surface.

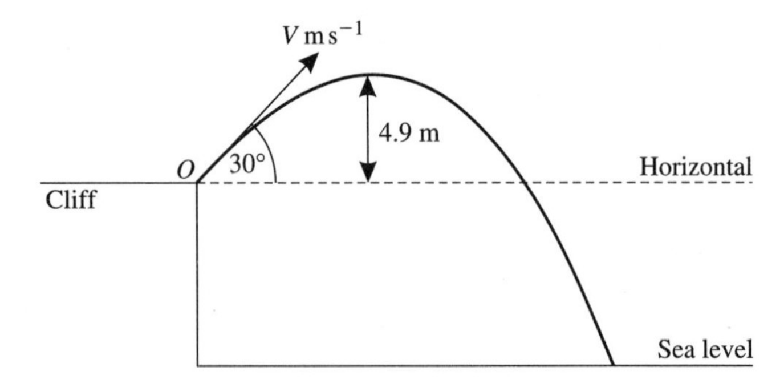

Consider the vertical motion to the highest point.

At this point, $s = 4.9\text{ m}$.

The initial vertical speed is $V\sin 30\text{ m s}^{-1} = \frac{1}{2}V\text{ m s}^{-1}$.

At the highest point $v = 0$.

Using: $v^2 = u^2 + 2as$ with $a = -9.8\text{ m s}^{-2}$

Gives:

$$0 = (\tfrac{1}{2}V)^2 + 2 \times (-9.8) \times 4.9$$

$$(\tfrac{1}{2}V)^2 = 2 \times 9.8 \times 4.9$$

$$\tfrac{1}{2}V = 9.8$$

$$V = 19.6$$

Consider now the vertical motion to sea level.

At sea level, the ball is 60 m below 0, so s = −60 m.

The initial vertical speed is: $\frac{1}{2}V\,\text{m s}^{-1} = 9.8\,\text{m s}^{-1}$

Using: $s = ut + \frac{1}{2}at^2$ with a $= -9.8\,\text{m s}^{-2}$

gives: $-60 = 9.8t - \frac{1}{2} \times 9.8t^2$

$$4.9t^2 - 9.8t - 60 = 0$$

This quadratic does not factorise so use the quadratic formula as shown in Book P1, chapter 2 to obtain

$$t = \frac{9.8 \pm \sqrt{(9.8^2 + 4 \times 4.9 \times 60)}}{2 \times 4.9}$$

$$t = \frac{9.8 \pm 35.66}{9.8}$$

The solution $t = \dfrac{9.8 - 35.66}{9.8}$ gives t negative which is not possible.

The other solution is $t = \dfrac{9.8 + 35.66}{9.8} = 4.639\cdot$

The time taken to reach the sea is 4.64 s.

Exercise 1B

Whenever a numerical value of g is required take $g = 9.8\,\text{m s}^{-2}$.

1 A particle is projected with a speed of $49\,\text{m s}^{-1}$ at an angle of $45°$ above the horizontal. Find (a) the time taken by the particle to reach its maximum height (b) the maximum height reached (c) the time of flight (d) the horizontal range of the particle.

2 A particle is projected with a speed of $56\,\text{m s}^{-1}$ at an angle of $30°$ above the horizontal. Find (a) the time taken by the particle to reach its maximum height (b) the maximum height reached (c) the time of flight (d) the horizontal range of the particle.

3 A particle is projected with a speed of $28\,\text{m s}^{-1}$ at an angle α above the horizontal. If the greatest height reached above the point of projection is 14 m find the value of α.

4 A particle is projected with a speed of $21\,\text{m s}^{-1}$ at an angle of elevation of $60°$. Find its speed and direction of motion after (a) 1 second (b) 2 seconds (c) 3 seconds.

5 A ball is thrown from O with a speed of $28\,\text{m}\,\text{s}^{-1}$ at an angle of elevation of 60°. It hits a wall which is 5 m horizontally from O. By modelling the ball as a particle calculate the height above O at which the ball hits the wall.

6 A ball is thrown at $14\,\text{m}\,\text{s}^{-1}$ at an angle of elevation of 60°. Find its horizontal and vertical distances from the point of projection after 1 second. Find the direction of motion of the ball at this time.

7 A particle is projected from the origin at a velocity of $(5\mathbf{i} + 16\mathbf{j})\,\text{m}\,\text{s}^{-1}$ where $\mathbf{i}$ and $\mathbf{j}$ are unit vectors in the horizontal and upward vertical directions respectively. Find the position vector of the particle after 2 seconds. Find its distance from the origin after 3 seconds.

8 A particle is projected from the origin at a velocity of $(4\mathbf{i} + 21\mathbf{j})\,\text{m}\,\text{s}^{-1}$ where $\mathbf{i}$ and $\mathbf{j}$ are unit vectors in the horizontal and upward vertical directions respectively. Find its position and velocity vectors after 2 seconds.

9 A golfer hits a golf ball resting on a tee with a velocity of $49\,\text{m}\,\text{s}^{-1}$ at an angle θ above the horizontal where $\sin\theta = \frac{3}{5}$. By modelling the ball as a particle and the ground as horizontal find the distance the ball travels before it hits the ground for the first time. What assumption have you made about the point of projection when using this model?

10 A stone is thrown at an angle of elevation of 30°. If 1 s later it hits the ground 1 m below its point of projection find the speed of projection. Find its greatest height above the point of projection.

11 A ball is thrown from O with a speed of $30\,\text{m}\,\text{s}^{-1}$ at an angle of 30° above the horizontal. Given that it just clears the top of a wall at a horizontal distance of 20 m from O, find the height of the top of the wall above O.

12 A gun fires a shell with an initial velocity of $770\,\text{m}\,\text{s}^{-1}$ at an angle of 18° above the horizontal. By modelling the shell as a particle find the range of the shell.

13 A stone is thrown from the top of a cliff which is 80 m above sea level. Initially the stone is moving at $20\,\text{m}\,\text{s}^{-1}$ at an angle of elevation of $\arcsin\frac{3}{5}$. Find the time taken by the stone to reach the surface of the sea and the horizontal distance travelled in this time.

14 A ball is hit by a racquet which is almost touching the ground. Two seconds later the ball just clears a vertical wall 4 m away horizontally and 4 m high. By modelling the ball as a particle projected at ground level and the ground as horizontal, find the range of the ball and its initial direction of motion.

15 An aeroplane is climbing at an angle of 2° while maintaining a speed of 400 m s^{-1}. A package is released and travels a horizontal distance of 2500 m before hitting the ground. By modelling the package as a particle projected with initial velocity the same as the velocity of the aeroplane, find the height of the aeroplane above the ground at the moment the package was released.

16 A child throws a ball with a speed of 20 m s^{-1} from a window 5 m above horizontal ground. If the ball hits the ground 1.5 seconds later find the direction of projection.

17 A stone is projected from a point O on a cliff with a speed of 20 m s^{-1} at an angle of elevation of 30°. T seconds later the angle of depression of the stone from O is 45°. Find the value of T.

18 A particle is projected from a point with position vector $(3\mathbf{i} + 3\mathbf{j})$ m relative to a fixed origin O where $\mathbf{i}$ and $\mathbf{j}$ are unit vectors in the horizontal and upward vertical directions respectively. Two seconds later the particle passes through the point with position vector $(13\mathbf{i} + 8\mathbf{j})$ m. Find its initial velocity vector.

19 A particle is projected with velocity vector $(10\mathbf{i} + 15\mathbf{j})$ m s^{-1} where $\mathbf{i}$ and $\mathbf{j}$ are unit vectors in the horizontal and upward vertical directions respectively. Find its velocity vector 2 seconds later and its distance from the point of projection at this time.

20 A particle is projected from a point on level ground. Its maximum height is 20 m and it hits the ground 200 m from its point of projection. Find the time of flight and the angle of projection.

21 A ball thrown at an angle arcsin $\frac{4}{5}$ to the horizontal just clears the top of a wall 60 m away horizontally 4 s after projection. By modelling the ball as a particle find (a) the velocity of projection (b) the height of the wall (c) the distance beyond the wall at which the ball hits the ground for the first time.

22 A particle is projected from O with a speed of $50\,\text{m s}^{-1}$ at an angle of 30° above the horizontal. Find the greatest height reached above the horizontal plane through O. Find the length of time for which the particle is more than 20 m above the plane. The particle reaches the horizontal plane through O again at P. Find the distance OP.

1.2 Velocity and acceleration when displacement is a function of time

In Book M1 you learnt about the motion of a particle moving in a straight line with constant acceleration. However, the acceleration of a moving particle is not always constant. If the acceleration is variable then calculus must be used to find the velocity of and the distance travelled by a particle. When a particle is travelling in a straight line and the displacement x measured from a fixed point on that line is given by:

$$x = \mathrm{f}(t)$$

the velocity, which is the rate of change of the displacement with respect to time, is given by:

$$v = \frac{\mathrm{d}x}{\mathrm{d}t}$$

Acceleration is the rate of change of velocity with respect to time and is given by:

$$a = \frac{\mathrm{d}v}{\mathrm{d}t}$$

Substituting $v = \dfrac{\mathrm{d}x}{\mathrm{d}t}$ gives:

$$a = \frac{\mathrm{d}^2x}{\mathrm{d}t^2}$$

Example 8

A particle P moves in a straight line such that its displacement x m from a fixed point O at time t seconds is given by $x = 2 + 3t - 2t^3$. Find:

(a) the distance of P from O when $t = 0$
(b) the distance of P from O when $t = 2$
(c) the velocity of P when $t = 0$
(d) the acceleration when $t = 1$.

(a) You know that: $x = 2 + 3t - 2t^3$

Substituting $t = 0$ in this expression gives $x = 2$.

When $t = 0$ the particle is 2 m from O.

(b) Substituting $t = 2$ gives:

$$x = 2 + (3 \times 2) - (2 \times 2^3) = 2 + 6 - 16 = -8$$

The **displacement** of the particle from O is -8 m when $t = 2$.
This means that the distance of the particle from O is 8 m but it is now on the opposite side of O from its position when $t = 0$.

(c) Using: $x = 2 + 3t - 2t^3$

Let the velocity of P be $v\,\text{m s}^{-1}$.
Differentiating x with respect to t gives:

$$v = \frac{dx}{dt} = 3 - 6t^2$$

Substituting $t = 0$ gives $v = 3$.

So the velocity is $3\,\text{m s}^{-1}$ when $t = 0$.

(d) Using: $v = 3 - 6t^2$

If the acceleration is $a\,\text{m s}^{-2}$ then:

$$a = \frac{dv}{dt} = -12t$$

Substituting $t = 1$ gives $a = -12$.
The acceleration of the particle is $-12\,\text{m s}^{-2}$ when $t = 1$. It has a retardation of $12\,\text{m s}^{-2}$.

Determining the velocity and displacement when the acceleration or velocity is given

If the acceleration is given as a function of time, then you need to reverse the process to find the velocity and displacement. In other words, you need to use integration

Suppose at time t seconds the acceleration is a m s^{-2}, the velocity v m s^{-1} and the displacement x m.

Since: $$a = \frac{dv}{dt} = f(t)$$

Integrating with respect to time gives:

$$v = \int f(t)\, dt + c$$

where c is the constant of integration.

Similarly if: $v = g(t)$

then: $x = \int g(t)\, dt + k$

where k is the constant of integration.

Example 9

A particle P moves in a straight line so that at time t seconds its acceleration is $3t\,\text{m}\,\text{s}^{-2}$. Given that P passes through the point O at time $t = 0$ with velocity $2\,\text{m}\,\text{s}^{-1}$, find

(a) the velocity of P in terms of t

(b) the distance of P from O when $t = 3$.

(a) Since: $$a = \frac{\mathrm{d}v}{\mathrm{d}t} = 3t$$

Integrating with respect to time gives:

$$v = \tfrac{3}{2}t^2 + c$$

Since $t = 0$ when $v = 2$, substituting these values into the equation gives $c = 2$.

So the velocity of P is given by:

$$v = \tfrac{3}{2}t^2 + 2$$

(b) You now know that:

$$v = \frac{\mathrm{d}x}{\mathrm{d}t} = \tfrac{3}{2}t^2 + 2$$

Integrating with respect to time gives:

$$x = \tfrac{3}{2} \times \frac{t^3}{3} + 2t + k$$

$$= \frac{t^3}{2} + 2t + k$$

You know that when $t = 0$ $x = 0$.

Substituting these values gives $k = 0$.

So: $$x = \frac{t^3}{2} + 2t$$

You need to find x when $t = 3$.

Substituting $t = 3$ gives:

$$x = \tfrac{3^3}{2} + 2 \times 3$$

$$= 13\tfrac{1}{2} + 6$$

$$= 19\tfrac{1}{2}$$

The particle is $19\frac{1}{2}$ m from O when $t = 3$.

Example 10

A particle moves in a straight line so that at time t seconds its velocity is given by $(2t^2 - 5t + 3)\,\text{m}\,\text{s}^{-1}$. Initially its displacement from the fixed point O on the line is 5 m. Find:

(a) the times when the particle is at rest

(b) the displacement of the particle from O when $t = 3$.

(a) You know that $v = 2t^2 - 5t + 3$

The particle is at rest when $v = 0$.

Hence it is at rest when $0 = 2t^2 - 5t + 3$.

Factorising gives:

$$0 = (2t - 3)(t - 1)$$

So either: $t = \frac{3}{2}$

Or: $t = 1$

The particle is at rest when $t = 1$ s and $t = 1\frac{1}{2}$ s.

(b) The displacement of the particle can be found by using:

$$v = \frac{dx}{dt} = 2t^2 - 5t + 3$$

Integrating with respect to time t gives:

$$x = \frac{2}{3}t^3 - \frac{5}{2}t^2 + 3t + c$$

Initially that is when $t = 0$, the particle's displacement from the fixed point O on the line is 5 m.

Substituting $t = 0$, $x = 5$ gives $c = 5$.

So: $x = \frac{2}{3}t^3 - \frac{5}{2}t^2 + 3t + 5$

Substituting $t = 3$ gives: $x = \left(\frac{2}{3} \times 3^3\right) - \left(\frac{5}{2} \times 3^2\right) + (3 \times 3) + 5$

$$= 18 - 22\tfrac{1}{2} + 9 + 5 = 9\tfrac{1}{2}$$

The particle is $9\frac{1}{2}$ m from O when $t = 3$.

Exercise 1C

1 A particle moves in a straight line such that at time t its displacement x from a fixed point O on that line is given by $x = 3t^3 - 2t^2$. Find (a) its velocity when $t = 2$ and (b) its acceleration when $t = 1$.

2 A particle moves in a straight line such that at time t its displacement x from a fixed point O on that line is given by $x = 3t^3 - 2t^2$. Find (a) the times when it is at rest and (b) its distances from O when it is at rest.

3 A particle moves in a straight line with velocity v given by $v = 3t - 4t^3 + 1$ at time t s after passing through a fixed point O in that line. Find (a) its velocity when $t = 1$ (b) its acceleration when $t = 2$ (c) its maximum velocity (d) its distance from O when $t = 2$.

4 A particle P moves in straight line such that at time t s its acceleration is given by $a = 2t - 2$. It passes through a point O on the line when $t = 0$ with a velocity of $1\,\text{m s}^{-1}$. Find (a) the velocity of P when $t = 2$ (b) the displacement of P from O when $t = 2$ (c) the time when P is at rest (d) the distance of P from O when it is at rest.

5 A particle P is moving along the x-axis with velocity $v = 5t^2 - 3t$. When $t = 0$ $x = 4$. Find (a) the times when P is at rest (b) the position of P each time it is at rest (c) the acceleration of P when $t = 2$.

6 A particle P is moving along the x-axis with velocity $v = 4t - 2t^2$. When $t = 0$ P is at $x = 3$. Find (a) the position of P when $t = 2$ (b) the maximum velocity attained by P (c) the distance OP when the velocity is maximum.

7 A particle P is moving along the x-axis. When $t = 0$ the velocity of P is $4\frac{1}{2}\,\text{m s}^{-1}$. At time t s the acceleration of P is given by $(3t - 6)\,\text{m s}^{-2}$. Find the time when the particle returns to its starting point.

8 A particle moves along the x-axis with velocity at time t given by $v = 5t^2 + 2t$. Find (a) the distance it moves in the 2nd second (b) the distance it moves in the 4th second.

1.3 Differentiating and integrating vectors

In chapter 2 of Book M1 you learnt about the application of vectors to displacements, velocities and accelerations when the velocity or acceleration was constant. In section 1.2 of this chapter you learnt to solve problems involving displacements, velocities and accelerations which were given as a function of time. In order to solve similar problems when they are given as vector functions of time you must first learn how to differentiate and integrate vectors.

Differentiating vectors

Mathematicians frequently use a dot as a quick way to denote differentiation with respect to time. With this notation:

$$\frac{d\mathbf{r}}{dt} = \dot{\mathbf{r}} \quad \text{and} \quad \frac{d^2\mathbf{r}}{dt^2} = \ddot{\mathbf{r}}$$

(The number of dots denotes the number of differentiations to be carried out.)

The basic rules of differentiation developed in Book P1, chapter 6, hold for vectors. So:

$$\frac{d}{dt}(\mathbf{r}_1 + \mathbf{r}_2) = \frac{d\mathbf{r}_1}{dt} + \frac{d\mathbf{r}_2}{dt}$$

and if $\mathbf{r} = f(t)\mathbf{c}$, where $\mathbf{c}$ is a constant vector, then:

$$\frac{d\mathbf{r}}{dt} = \frac{d}{dt}[f(t)\mathbf{c}] = \frac{df}{dt}\mathbf{c}$$

The unit vectors **i** and **j** are constant vectors, so if **r** is given in the **i**, **j** notation then:

$$\mathbf{r} = x\mathbf{i} + y\mathbf{j}$$
$$\dot{\mathbf{r}} = \dot{x}\mathbf{i} + \dot{y}\mathbf{j}$$
$$\ddot{\mathbf{r}} = \ddot{x}\mathbf{i} + \ddot{y}\mathbf{j}$$

Example 11

The position vector of a particle P at time t s is $\mathbf{r} = (2t - 1)\mathbf{i} + t^2\mathbf{j}$, where **r** is measured in metres. Find the initial position vector and velocity of P and show that the acceleration of P is constant.

When $t = 0$: $\mathbf{r} = -\mathbf{i}$

Therefore the initial position vector of P is $-\mathbf{i}$ m.

Differentiating with respect to t gives:

$$\dot{\mathbf{r}} = 2\mathbf{i} + 2t\mathbf{j}$$

So when $t = 0$: $\dot{\mathbf{r}} = 2\mathbf{i}\,\text{m s}^{-1}$

Differentiating again gives:

$$\ddot{\mathbf{r}} = 2\mathbf{j}\,\text{m s}^{-2}$$

So the acceleration is a constant vector.

Integrating vectors

The integration of vectors written in the **i**, **j** notation follows the rules of integration given in Book P1, chapter 7.

If: $\ddot{\mathbf{r}} = f(t)\mathbf{i} + g(t)\mathbf{j}$

then:

$$\dot{r} = \int \ddot{\mathbf{r}}\,dt$$
$$= \mathbf{i}\int f(t)\,dt + \mathbf{j}\int g(t)\,dt + C_1\mathbf{i} + C_2\mathbf{j}$$

- **When you integrate a vector you must add an arbitrary constant vector $C_1\mathbf{i} + C_2\mathbf{j}$, where C_1 and C_2 are constants.**

This is just a generalisation of the idea of adding an arbitrary constant when you integrate a function.

Similarly, if: $\dot{\mathbf{r}} = \mathrm{F}(t)\mathbf{i} + \mathrm{G}(t)\mathbf{j}$

then: $\mathbf{r} = \mathbf{i}\int \mathrm{F}(t)\,\mathrm{d}t + \mathbf{j}\int \mathrm{G}(t)\,\mathrm{d}t + k_1\mathbf{i} + k_2\mathbf{j}$

where k_1 and k_2 are constants.

Example 12

A particle P moves so that at time t:

$$\dot{\mathbf{r}} = 2t\mathbf{i} + 3t^2\mathbf{j}$$

When $t = 0$ its position vector is $4\mathbf{i} - 3\mathbf{j}$. Find the position vector $\mathbf{r}$ of P at time t.

As: $\dot{\mathbf{r}} = 2t\mathbf{i} + 3t^2\mathbf{j}$

then: $\mathbf{r} = \mathbf{i}\int 2t\,\mathrm{d}t + \mathbf{j}\int 3t^2\,\mathrm{d}t + k_1\mathbf{i} + k_2\mathbf{j}$

$= t^2\mathbf{i} + t^3\mathbf{j} + k_1\mathbf{i} + k_2\mathbf{j}$

At $t = 0$ this gives:

$$\mathbf{r} = 0\mathbf{i} + 0\mathbf{j} + k_1\mathbf{i} + k_2\mathbf{j}$$

But this must be equal to $4\mathbf{i} - 3\mathbf{j}$.

Equating $\mathbf{i}$ parts gives: $k_1 = 4$

Equating $\mathbf{j}$ parts gives: $k_2 = -3$

So: $\mathbf{r} = t^2\mathbf{i} + t^3\mathbf{j} + 4\mathbf{i} - 3\mathbf{j}$

$= (t^2 + 4)\mathbf{i} + (t^3 - 3)\mathbf{j}$

The position vector of P at time t is $\mathbf{r} = (t^2 + 4)\mathbf{i} + (t^3 - 3)\mathbf{j}$.

Example 13

A particle P moves so that $\ddot{\mathbf{r}} = p\mathbf{i} + q\mathbf{j}$ where p and q are constants. (This means $\ddot{\mathbf{r}}$ is a constant vector.) When $t = 0$, $\dot{\mathbf{r}} = \mathbf{0}$ and when $t = 1$, $\dot{\mathbf{r}} = 3\mathbf{i} - 2\mathbf{j}$. Find:

(a) $\dot{\mathbf{r}}$ at time t
(b) $|\dot{\mathbf{r}}|$ when $t = 3$.

(a) You know that: $\ddot{\mathbf{r}} = p\mathbf{i} + q\mathbf{j}$

So: $\dot{\mathbf{r}} = \mathbf{i}\int p\,\mathrm{d}t + \mathbf{j}\int q\,\mathrm{d}t + C_1\mathbf{i} + C_2\mathbf{j}$

$= pt\mathbf{i} + qt\mathbf{j} + C_1\mathbf{i} + C_2\mathbf{j}$

Since: $\dot{\mathbf{r}} = \mathbf{0}$ when $t = 0$:

$$\mathbf{0} = 0\mathbf{i} + 0\mathbf{j} + C_1\mathbf{i} + C_2\mathbf{j}$$

and so: $C_1 = 0$ and $C_2 = 0$

So: $\dot{\mathbf{r}} = pt\mathbf{i} + qt\mathbf{j}$

When $t = 1$ you know that $\dot{\mathbf{r}} = 3\mathbf{i} - 2\mathbf{j}$.

So: $p\mathbf{i} + q\mathbf{j} = 3\mathbf{i} - 2\mathbf{j}$

$$p = 3$$

$$q = -2$$

Hence: $\dot{\mathbf{r}} = 3t\mathbf{i} - 2t\mathbf{j}$

(b) When $t = 3$: $\dot{\mathbf{r}} = 3(3)\mathbf{i} - 2(3)\mathbf{j}$

$$= 9\mathbf{i} - 6\mathbf{j}$$

and $|\dot{\mathbf{r}}| = \sqrt{(9^2 + 6^2)}$

$$= \sqrt{(81 + 36)}$$

$$= \sqrt{117} = 10.8$$

Example 14

At time $t = 0$ a particle P is at the point with position vector $(30\mathbf{i} + 35\mathbf{j})$ m relative to an origin O. P moves so that $\ddot{\mathbf{r}}$ is constant and equal to $(3\mathbf{i} + 4\mathbf{j})\,\text{m s}^{-2}$. When $t = 0$, $\dot{\mathbf{r}} = \mathbf{0}$. Find:

(a) the vector $\dot{\mathbf{r}}$ when $t = 4$
(b) the distance of P from O at this time.

(a) As $\ddot{\mathbf{r}} = 3\mathbf{i} + 4\mathbf{j}$, integrating with respect to t gives:

$$\dot{\mathbf{r}} = \int \ddot{\mathbf{r}}\,\mathrm{d}t = \mathbf{i}\int 3\,\mathrm{d}t + \mathbf{j}\int 4\,\mathrm{d}t + C_1\mathbf{i} + C_2\mathbf{j}$$

$$= 3t\mathbf{i} + 4t\mathbf{j} + C_1\mathbf{i} + C_2\mathbf{j}$$

Using $\dot{\mathbf{r}} = \mathbf{0}$ when $t = 0$ gives:

$$\mathbf{0} = C_1\mathbf{i} + C_2\mathbf{j}$$

and so: $C_1 = 0$ and $C_2 = 0$

Hence: $\dot{\mathbf{r}} = 3\text{t}\mathbf{i} + 4\text{t}\mathbf{j}$

Substituting $t = 4$ in this result gives $\dot{\mathbf{r}}$ when $t = 4$ as:

$$\dot{\mathbf{r}} = (12\mathbf{i} + 16\mathbf{j})\,\text{m s}^{-1}$$

(b) The distance of P from O is just the magnitude of $\mathbf{r}$. So you must first find $\mathbf{r}$.

As $\dot{\mathbf{r}} = 3t\mathbf{i} + 4t\mathbf{j}$, integrating with respect to t gives:

$$\mathbf{r} = \int \dot{\mathbf{r}}\,\mathrm{d}t$$

$$= \mathbf{i}\int 3t\,\mathrm{d}t + \mathbf{j}\int 4t\,\mathrm{d}t + k_1\mathbf{i} + k_2\mathbf{j}$$

$$= \mathbf{i}\frac{3t^2}{2} + \mathbf{j}\frac{4t^2}{2} + k_1\mathbf{i} + k_2\mathbf{j}$$

When $t = 0$, $\mathbf{r} = 30\mathbf{i} + 35\mathbf{j}$, and so:

$$30\mathbf{i} + 35\mathbf{j} = 0\mathbf{i} + 0\mathbf{j} + k_1\mathbf{i} + k_2\mathbf{j}$$

Equating **i** and **j** components gives:

$$k_1 = 30, k_2 = 35$$

So: $$\mathbf{r} = \frac{3t^2}{2}\mathbf{i} + 2t^2\mathbf{j} + 30\mathbf{i} + 35\mathbf{j}$$

When $t = 4$: $$\mathbf{r} = 24\mathbf{i} + 32\mathbf{j} + 30\mathbf{i} + 35\mathbf{j}$$
$$= 54\mathbf{i} + 67\mathbf{j}$$

The magnitude of **r** is:

$$|\mathbf{r}| = \sqrt{(54^2 + 67^2)}$$
$$= 86.1$$

So the distance of P from O at this time is 86.1 m.

Motion of a particle in a plane

Studying the motion of a particle in two dimensions in the (x, y) plane involves using vectors. If unit vectors **i** and **j** are taken in the directions of the x-and y-axes respectively then the position vector **r** m relative to a fixed origin O in the plane of a moving particle P may be given by:

$$\mathbf{r} = \mathrm{f}(t)\mathbf{i} + \mathrm{g}(t)\mathbf{j}$$

where f(t) and g(t) are both functions of time t.

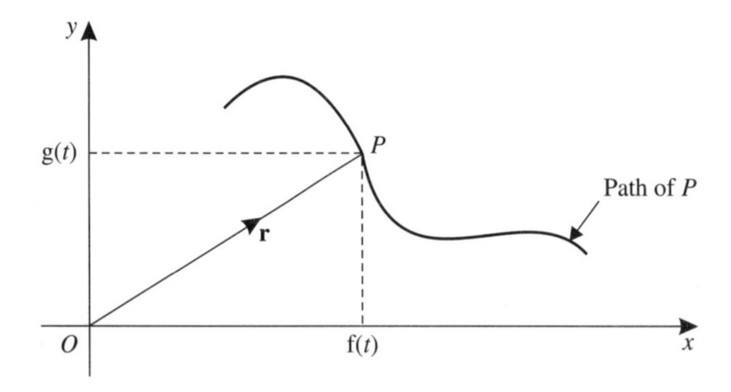

The velocity and acceleration of P can be found by differentiating each component separately, as shown before. With the usual notation of velocity being $\mathbf{v}\,\mathrm{m\,s^{-1}}$ and acceleration $\mathbf{a}\,\mathrm{m\,s^{-2}}$:

$$\mathbf{v} = \frac{\mathrm{d}\mathbf{r}}{\mathrm{d}t} \quad \text{and} \quad \mathbf{a} = \frac{\mathrm{d}\mathbf{v}}{\mathrm{d}t} = \frac{\mathrm{d}^2\mathbf{r}}{\mathrm{d}t^2}$$

As: $\mathbf{r} = f(t)\,\mathbf{i} + g(t)\mathbf{i}$

it follows that: $\mathbf{v} = \dfrac{df(t)}{dt}\mathbf{i} + \dfrac{dg(t)}{dt}\mathbf{j}$

and: $\mathbf{a} = \dfrac{d^2f(t)}{dt^2}\mathbf{i} + \dfrac{d^2g(t)}{dt^2}\mathbf{j}$

When integrating include the constant of integration in vector form.

Example 15

The position vector **r** metres of a particle P relative to a fixed point O is given by $\mathbf{r} = (5t^2 - 2t)\mathbf{i} + (t - 3)\mathbf{j}$. Find the velocity and acceleration in vector form.

As: $\mathbf{r} = (5t^2 - 2t)\mathbf{i} + (t - 3)\mathbf{j}$

it follows that: $\mathbf{v} = \dfrac{d\mathbf{r}}{dt} = (10t - 2)\mathbf{i} + \mathbf{j}$

and: $\mathbf{a} = \dfrac{d\mathbf{v}}{dt} = 10\mathbf{i}$

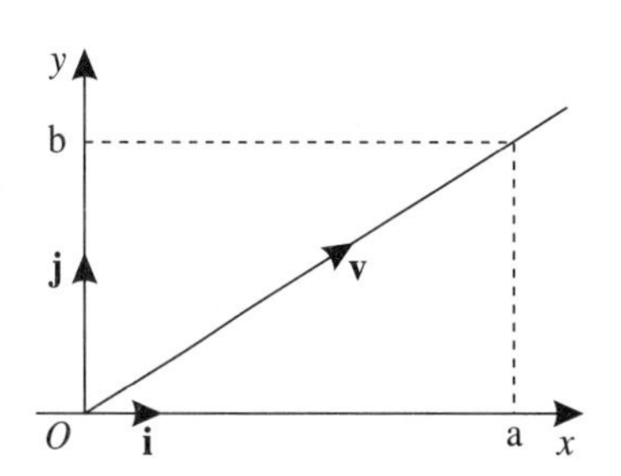

The velocity vector is $[(10t - 2)\mathbf{i} + \mathbf{j}]\,\mathrm{m\,s^{-1}}$ and the acceleration vector is $10\mathbf{i}\ \mathrm{m\,s^{-2}}$.

The speed of a particle is the **magnitude** of its velocity. (Speed is a scalar quantity not a vector.)

$$\text{speed} = |\text{velocity}|$$

If $\mathbf{v} = \mathrm{a}\mathbf{i} + \mathrm{b}\mathbf{j}$ where a and b are constants then by Pythagoras' Theorem $|\mathbf{v}| = \sqrt{(\mathrm{a}^2 + \mathrm{b}^2)}$.

■ $$\text{speed} = \sqrt{(\mathrm{a}^2 + \mathrm{b}^2)}\ \mathrm{m\,s^{-1}}$$

Example 16

A particle P moves so that its acceleration $\mathbf{a}\,\mathrm{m\,s^{-2}}$ at time t seconds is given by $\mathbf{a} = 2t\mathbf{i} + 4\mathbf{j}$. When $t = 0$ the velocity of P is $(3\mathbf{i} - 7\mathbf{j})\,\mathrm{m\,s^{-1}}$. Find the speed of P when $t = 3$.

$$\mathbf{a} = \frac{d\mathbf{v}}{dt} = 2t\mathbf{i} + 4\mathbf{j}$$

By integration $\mathbf{v} = t^2\mathbf{i} + 4t\mathbf{j} + \mathrm{A}\mathbf{i} + \mathrm{B}\mathbf{j}$ where A and B are scalar constants.

When $t = 0$: $\mathbf{v} = 3\mathbf{i} - 7\mathbf{j}$

So: $3\mathbf{i} - 7\mathbf{j} = A\mathbf{i} + B\mathbf{j}$

Hence: $\mathbf{v} = t^2\mathbf{i} + 4t\mathbf{j} + 3\mathbf{i} - 7\mathbf{j}$

$\mathbf{v} = (t^2 + 3)\mathbf{i} + (4t - 7)\mathbf{j}$

Substituting $t = 3$ gives: $\mathbf{v} = (3^2 + 3)\mathbf{i} + (4 \times 3 - 7)\mathbf{j}$

$= 12\mathbf{i} + 5\mathbf{j}$

$$|\mathbf{v}| = \sqrt{(12^2 + 5^2)} = 13$$

The particle's speed when $t = 3$ is $13\,\mathrm{m\,s^{-1}}$.

Exercise 1D

In these questions, **i** and **j** are unit vectors in the direction of the positive x- and y-axes, respectively. The units of measure in questions 1 to 8 are metres and seconds.

1 The position of a particle P at time t is determined by its position vector $\mathbf{r}$, from which the velocity vector can be found. The diagram shows the position of P and the direction of its velocity when $\mathbf{r} = t\mathbf{i} + t^2\mathbf{j}$, and $t = 1$.

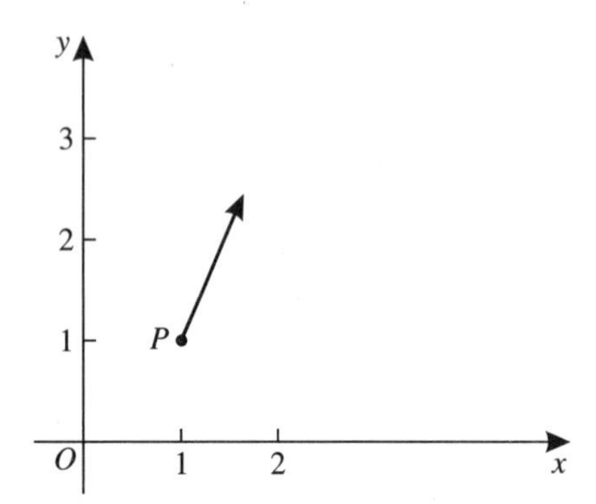

Draw similar diagrams showing the position and direction of motion of the particle when $t = 0$, $t = 1$ and $t = 2$ in each of the following cases:

(a) $\mathbf{r} = 2\mathbf{i} + t\mathbf{j}$ (b) $\mathbf{r} = 2\mathbf{i} + t^2\mathbf{j}$ c) $\mathbf{r} = t\mathbf{i} + t\mathbf{j}$
(d) $\mathbf{r} = t\mathbf{i} + 2\mathbf{j}$ (e) $\mathbf{r} = t^2\mathbf{i} + \frac{1}{2}t^3\mathbf{j}$.

2 Find the velocity vector $\mathbf{v}$ for each part of question 1, and the velocity and speed in each case when $t = 1$.

3 A particle P has position vector $\mathbf{r}$. Find $\dot{\mathbf{r}}$ and $\ddot{\mathbf{r}}$ in each of the following cases:

(a) $\mathbf{r} = t\mathbf{i} + t^2\mathbf{j}$ (b) $\mathbf{r} = \frac{1}{2}t^2\mathbf{i} + t\mathbf{j}$ (c) $\mathbf{r} = t^3\mathbf{i} + t\mathbf{j}$
(d) $\mathbf{r} = t^3\mathbf{i} + \mathbf{j}$ (e) $\mathbf{r} = \frac{1}{3}t^3\mathbf{i} + \frac{1}{4}t^4\mathbf{j}$.

4 A particle P has position vector $\mathbf{r}$. Find the velocity and acceleration vectors in each of the following cases:

(a) $\mathbf{r} = t^2\mathbf{i} + \mathbf{j}$ (b) $\mathbf{r} = t^2\mathbf{i} + t\mathbf{j}$ (c) $\mathbf{r} = t^2\mathbf{i} + t^2\mathbf{j}$
(d) $\mathbf{r} = t^3\mathbf{i} + t^3\mathbf{j}$ (e) $\mathbf{r} = \frac{1}{4}t^4\mathbf{i} + \frac{1}{3}t^3\mathbf{j}$.

5 For each part of question 4 find the speed of the particle when $t = 2$.

6 A particle P is initially at the origin and has velocity vector $\mathbf{v}$. Find the position vector $\mathbf{r}$ and the acceleration vector $\mathbf{a}$ in each of the following cases:
(a) $\mathbf{v} = 2t\mathbf{i} + 2\mathbf{j}$ (b) $\mathbf{v} = 2t\mathbf{i} + 3t^2\mathbf{j}$ (c) $\mathbf{v} = \mathbf{i} + 3t^2\mathbf{j}$
(d) $\mathbf{v} = 4t^3\mathbf{i} + 3t^2\mathbf{j}$ (e) $\mathbf{v} = t^2\mathbf{i} + 12t^3\mathbf{j}$.

7 A particle P is initially at the origin O with velocity $\mathbf{i}$. Find the velocity vector $\mathbf{v}$ and the position vector $\mathbf{r}$ in each of the following cases, where $\mathbf{a}$ is the acceleration vector:
(a) $\mathbf{a} = \mathbf{i}$ (b) $\mathbf{a} = \mathbf{j}$ (c) $\mathbf{a} = \mathbf{i} + \mathbf{j}$
(d) $\mathbf{a} = 6t\mathbf{i} + \mathbf{j}$ (e) $\mathbf{a} = \mathbf{i} + 6t\mathbf{j}$.

8 Sketch diagrams similar to those in question 1 to show the path of the particle P for the motion described in question 4(a), (b), (c).

9 The position vector $\mathbf{r}$ m of a particle P relative to a fixed point O is given by $\mathbf{r} = (3t^2 - 4t)\mathbf{i} + (2t^2 - 1)\mathbf{j}$. Find the velocity vector of P and show that the acceleration of P is constant.

10 A particle P is moving in the (x, y) plane with velocity $\mathbf{v}\,\mathrm{m\,s^{-1}}$ at time t seconds given by $\mathbf{v} = 2t^2\mathbf{i} + 2t^{\frac{3}{2}}\mathbf{j}$.
(a) Find the acceleration of P when $t = 4$.
Given that P is at O when $t = 0$:
(b) Find the position vector of P relative to O when $t = 1$
(c) Find the distance of P from O when $t = 1$.

11 A particle P starts from rest at a fixed origin O when $t = 0$. The acceleration $\mathbf{a}\,\mathrm{m\,s^{-2}}$ of P at time t seconds is given by $\mathbf{a} = 6t\mathbf{i} - 4\mathbf{j}$. Find:
(a) the velocity vector of P when $t = 2$
(b) the speed of P when $t = 2$
(c) the position vectors of P relative to O when $t = 1$ and $t = 2$
(d) the distance between the positions of P when $t = 1$ and $t = 2$.

12 A particle moves with acceleration $\mathbf{a}\,\mathrm{m\,s^{-2}}$ at time t seconds given by $\mathbf{a} = 7t\mathbf{i} + 4t^2\mathbf{j}$. Given that $\mathbf{v} = 3\mathbf{i} - 4\mathbf{j}$ when $t = 0$ find the speed of the particle when $t = 2$.

13 When $t = 0$ a particle P is at rest at point A whose position vector relative to a fixed point O is $\mathbf{r} = (3\mathbf{i} + 4\mathbf{j})$ m. P moves with acceleration $\mathbf{a}\,\mathrm{m\,s^{-2}}$ at time t seconds given by $\mathbf{a} = 5t\mathbf{i}$. Find the distance P moves between $t = 1$ and $t = 2$.

14 When $t = 0$ a particle P is at the point with position vector $(3\mathbf{i} + 4\,\mathbf{j})$ m relative to a fixed point O. The velocity $\mathbf{v}$ m s^{-1} at time t seconds of P is given by $\mathbf{v} = 6t^2\mathbf{i} + (2t + 3)\mathbf{j}$.
(a) Find the position vector of P relative to O at time t seconds. When $t = 0$ a second particle Q is at the point with position vector $(2\mathbf{i} + 10\mathbf{j})$ m relative to O and the velocity $\mathbf{v}$ m s^{-1} at time t seconds for Q is given by $\mathbf{v} = 3\mathbf{i} - 2\mathbf{j}$.
(b) Find the position vector of Q relative to O at time t seconds.
(c) Show that P and Q will collide when $t = 1$ and find the position vector of their point of collision.

15 A particle P moves so that at time t seconds its position vector $\mathbf{r}$ m relative to a fixed origin O is given by:

$$\mathbf{r} = (2t^2 - \mathrm{p}t)\mathbf{i} + (t^3 - 3t)\mathbf{j}$$

where p is a constant.
(a) Find an expression for the velocity, $\mathbf{v}$ m s^{-1}, at time t seconds.
(b) Given that the particle comes to instantaneous rest find the time at which it is at rest and hence the value of p.

16 At time t seconds two particles A and B have position vectors $\mathbf{r}_\mathrm{a}$ m and $\mathbf{r}_\mathrm{b}$ m respectively with respect to a fixed origin O where:

$$\mathbf{r}_\mathrm{a} = (t^2 + 12t)\mathbf{i} + 4t\mathbf{j}$$
$$\mathbf{r}_\mathrm{b} = t^2\mathbf{i} + (2t - 6)\mathbf{j}$$

Find the velocity vectors $\mathbf{v}_\mathrm{a}$ m s^{-1} and $\mathbf{v}_\mathrm{b}$ m s^{-1} of A and B at time t seconds.
Hence determine the value of t for which the speed of A is twice the speed of B.

17 At time t seconds two particles P and Q have position vectors $\mathbf{r}_\mathrm{p}$ m and $\mathbf{r}_\mathrm{q}$ m respectively relative to a fixed origin O where:

$$\mathbf{r}_\mathrm{p} = \tfrac{1}{2}t^2\mathbf{i} + 2\mathbf{j} \quad \text{and} \quad \mathbf{r}_\mathrm{q} = 2t(3\mathbf{i} - 4\mathbf{j})$$

(a) Calculate the distance between P and Q when $t = 2$
(b) Show that the velocity of Q is constant and calculate its magnitude and direction
(c) Calculate the acceleration of P.

Exercise 1E Mixed questions

Whenever a numerical value of g is required take $g = 9.8\,\text{m s}^{-2}$.

1 A particle moves along the x-axis, passing through the origin O with speed $3\,\text{m s}^{-1}$ in the positive direction. At time t seconds after passing through O the acceleration of the particle is $(2t - 1)\,\text{m s}^{-2}$.

(a) Find the speed of the particle when $t = 4$.

(b) Find the distance of the particle from O at time t seconds.

(c) Hence find the distance travelled by the particle between the instants when $t = 1$ and $t = 3$.

2 A cricket ball is hit from a point which is 1.2 m above a horizontal pitch. Initially the ball is moving with a speed of $16.7\,\text{m s}^{-1}$ at an angle of elevation of $35°$. By modelling the ball as a particle moving freely under gravity, find:

(a) the time that elapses before the ball hits the ground

(b) the horizontal distance from the point where the ball was hit to the point where it hits the ground.

(c) State a physical factor that you have ignored in your model.

3 At time t seconds, the acceleration $\mathbf{a}\,\text{m s}^{-2}$ of a particle P is given by

$$\mathbf{a} = (2t + 3)\mathbf{i} - 4t\mathbf{j}$$

where $\mathbf{i}$ and $\mathbf{j}$ are perpendicular horizontal vectors. P has mass 3 kg.

(a) Find, in newtons, the magnitude of the force acting on P when $t = 6$.

Given that P has velocity $(3\mathbf{i} + 2\mathbf{j})\,\text{m s}^{-1}$ when $t = 0$, find:

(b) the velocity of P at time t seconds

(c) the value of t when P is moving in a direction parallel to the vector $\mathbf{i}$.

4 A particle P of mass 1.5 kg moves under the action of a single force $\mathbf{F}$ newtons. The position vector $\mathbf{r}$ metres of P at time t seconds relative to a fixed origin O is given by

$$\mathbf{r} = (3t^2 + 4)\mathbf{i} + (4t - t^2)\mathbf{j}$$

where $\mathbf{i}$ and $\mathbf{j}$ are perpendicular unit vectors. Find:

(a) the speed of P when $t = 4$

(b) the angle between the direction of motion of P and the vector $\mathbf{i}$ when $t = 4$

(c) the magnitude of $\mathbf{F}$.

5 A particle P moves along the x-axis. P is initially at rest at the origin O. At time t seconds the acceleration of P in the direction of x increasing is $(3t^2 - 14t + 12)\,\text{m s}^{-2}$. Find:

(a) the non-zero values of t at which P is instantaneously at rest

(b) the distance travelled by P in the interval $1 \leqslant t \leqslant 2$.

6 A ball is hit from a height of 1.4 m above horizontal ground with a speed of $20\,\text{m s}^{-1}$ at an angle of elevation 25°. The ball just clears a vertical wall which is 30 m horizontally from the point where it was struck. By modelling the ball as a particle moving freely under gravity, find:

(a) the height of the wall

(b) the speed of the ball at the instant when it passes over the wall

(c) the direction of motion of the ball at the instant when it passes over the wall.

7 The velocity $\mathbf{v}\,\text{m s}^{-1}$ of a particle P at time t seconds is given by

$$\mathbf{v} = (3t^2 - 12)\mathbf{i} + 5t\mathbf{j}$$

(a) Find the acceleration of P at time t seconds.

(b) Find the value of t when P is moving parallel to the vector $\mathbf{j}$.

With respect to a fixed origin O, the position vector of P when $t = 2$ is $4\mathbf{j}$.

(c) Find the distance OP when $t = 1$.

8 A particle P of mass 0.5 kg is at rest on a horizontal table. It receives a blow of impulse 2.5 N s.

(a) Calculate the speed with which P is moving immediately after the blow.

The height of the table is 0.9 m and the floor is horizontal.
In an initial model of the situation the table is assumed to be smooth.
(b) Calculate the horizontal distance from the edge of the table to the point where P hits the ground.
In a refinement of the model the table is assumed to be rough.
The coefficient of friction between P and the table is 0.2.
(c) Calculate the deceleration of P.
Given that P travels 0.4 m to the edge of the table,
(d) calculate the time which elapses between P receiving the blow to P hitting the floor.

9 At times t seconds the position vector $\mathbf{r}$ metres of a particle, P, of mass 3 kg, is given by

$$\mathbf{r} = (3t^2 + 1)\mathbf{i} + (24t - 2t^3)\mathbf{j}, \ t > 0$$

(a) Find the speed, in m s^{-1}, of P when $t = 1$.
(b) Find the value of t when the velocity of P is perpendicular to $\mathbf{j}$.
(c) Find the magnitude and direction of the force acting on P when $t = 1$, giving the direction as an angle made with $\mathbf{i}$, to the nearest degree.

10

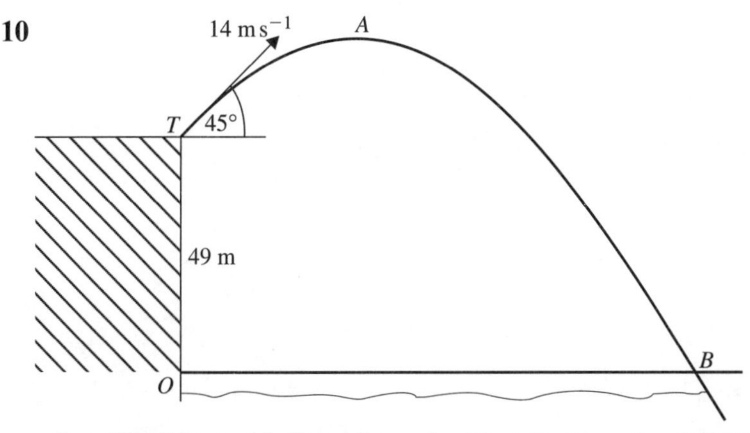

A golf ball is struck from the point T, at the top of a cliff 49 m above sea level, with a speed of 14 m s^{-1} at an angle of 45° to the horizontal, as shown in the diagram. The point O is at sea level and vertically below T. The point A is the highest point reached by the ball in its motion. The ball strikes the sea at the point B.
(a) Find the height A above sea level.
(b) Find the distance OB.
(c) Discuss briefly the assumptions you have made in your calculation and give reasons why they are justified in this case.

SUMMARY OF KEY POINTS

1 **Projectiles**
The horizontal speed of a projectile is unchanged throughout the motion. The vertical motion is subject to an acceleration of magnitude $g = 9.8\,\text{m}\,\text{s}^{-2}$ vertically downwards. The four uniform acceleration equations apply.

2 **Variable acceleration**
For variable acceleration the relationships between the displacement x, velocity v and acceleration a of a particle are shown by:

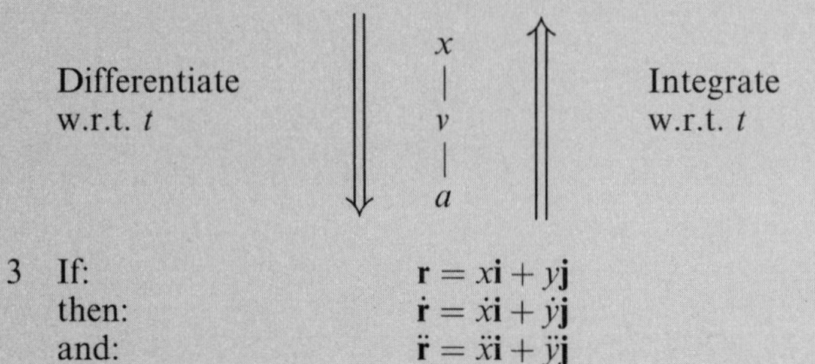

3 If: $\mathbf{r} = x\mathbf{i} + y\mathbf{j}$
then: $\dot{\mathbf{r}} = \dot{x}\mathbf{i} + \dot{y}\mathbf{j}$
and: $\ddot{\mathbf{r}} = \ddot{x}\mathbf{i} + \ddot{y}\mathbf{j}$

4 If the position vector $\mathbf{r}$ of a particle is given

then: $$\mathbf{v} = \dot{\mathbf{r}} = \frac{\mathrm{d}\mathbf{r}}{\mathrm{d}t}$$

and: $$\mathbf{a} = \dot{\mathbf{v}} = \ddot{\mathbf{r}} = \frac{\mathrm{d}\mathbf{v}}{\mathrm{d}t} = \frac{\mathrm{d}^2\mathbf{r}}{\mathrm{d}t^2}$$

5 The speed of a particle is the magnitude of its velocity vector, that is:

$$\text{speed} = |\mathbf{v}|$$

Centres of mass

2

In Book M1 you learnt about the moment of a force and how to solve simple problems using moments. In this chapter you are going to learn about another use of moments – finding the position of the centre of mass of a discrete mass distribution and of some uniform plane figures.

2.1 Centre of mass of a discrete mass distribution

Every particle in a discrete mass distribution has a force pulling the particle towards the centre of the earth. This force is called the weight of the particle. Since the centre of the earth is at a great distance from the earth's surface, you can assume that all these weights are pulling vertically downwards and that these forces are parallel to each other. The weight of the system is the resultant of the weights of the particles and this resultant force acts through a point called the **centre of mass** of the system.

Centre of mass of a system of particles distributed in one dimension

Consider a simple system consisting of two particles A and B of equal mass m kg. The centre of mass is at the mid-point of the line joining A and B.

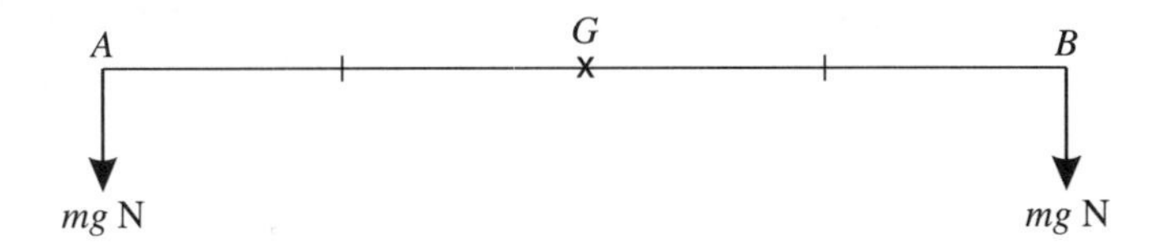

G is the centre of mass of A and B.

Where the particles have different masses or there are more than two particles involved, the method of calculating the centre of mass of the system is shown by example 1.

Example 1

Three particles of masses 5 kg, 3 kg and 4 kg are attached to a light rod AB of length 2 m at the points A, B and C where $AC = 0.6$ m. Find the position of the centre of mass of the system.

Assume the rod to be horizontal and start by drawing two diagrams, one showing the weights of the separate particles and the other showing the resultant weight.

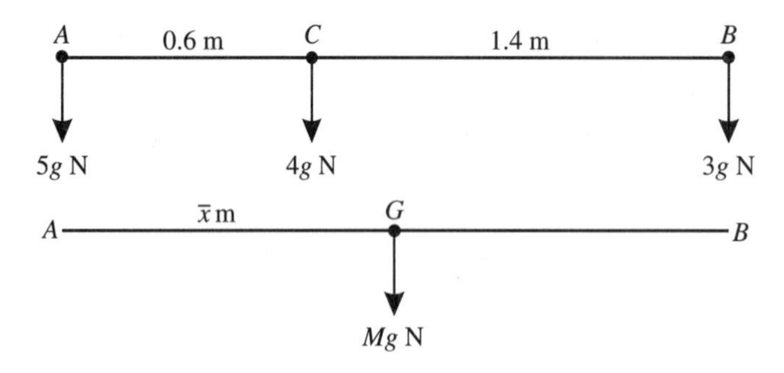

Let the total mass be M kg and the centre of mass of the system be at the point G where $AG = \bar{x}$ m. The forces in the two diagrams are equivalent. So resolving vertically will give the same result for each diagram.

Resolving vertically gives:

$$Mg = 5g + 4g + 3g$$

So :

$$M = 12$$

The total mass is 12 kg.

Similarly, both force systems will have the same total moment about any point.

Taking moments about point A gives:

M(A): $\quad$ @ $Mg\,\bar{x} = 4g \times 0.6 + 3g \times 2$

Substituting $M = 12$ gives:

$$12g\bar{x} = 2.4g + 6g$$

$$\bar{x} = \frac{8.4}{12} = 0.7$$

The centre of mass of the particles is 0.7 m from A.

This method can be applied to a larger system which has n particles of masses m_1, m_2, $\dots m_n$ arranged in a straight line where the particles $m_1, m_2, \dots m_n$ are at a distance $x_1, x_2, \dots x_n$ respectively from an origin O on the line.

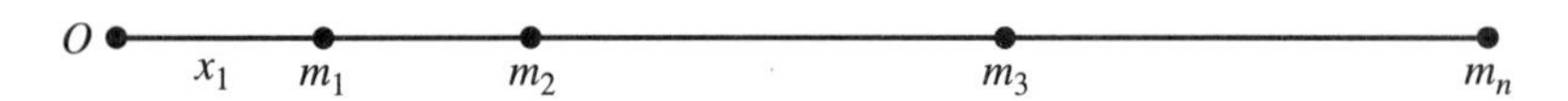

Let the centre of mass of the system be a distance $\bar{x}$ from O.

O ——— $\bar{x}$ ——— M ●———

In example 1, the weights of the particles were used in the moments equation. Because g was a factor of every term it could be cancelled from the equation. You can therefore work with the masses of the particles rather than the weights.

Then the total mass M for the above system is:

$$M = m_1 + m_2 + m_3 + \ldots + m_n$$

Or: $$M = \Sigma m_i$$

Taking moments about O for both diagrams gives:

$$M\bar{x} = m_1x_1 + m_2x_2 + m_3x_3 + \ldots + m_nx_n$$

Or: $$M\bar{x} = \Sigma m_ix_i$$

Since $M = \Sigma m_i$: $$\bar{x} = \frac{\Sigma m_ix_i}{\Sigma m_i}$$

Centre of mass of a system of particles distributed in two dimensions

If the particles in a system are not in a straight line you need to give their positions as coordinates relative to some fixed axes. The coordinates of the centre of mass can then be found by using the same method as for a one-dimensional distribution (example 1). This time the method must be applied twice, once using x-coordinates to give:

$$\bar{x} = \frac{\Sigma m_ix_i}{\Sigma m_i}$$

and again using y-coordinates to give:

$$\bar{y} = \frac{\Sigma m_iy_i}{\Sigma m_i}$$

where $(\bar{x}, \bar{y})$ are the coordinates of the centre of mass of the system.

In some questions axes and coordinates are specified, in others you need to choose your own axes.

Using a tabular form to display both the masses of the particles and their coordinates will help you form correct equations.

Example 2

Calculate the coordinates of the centre of mass of particles of mass 4 kg, 3 kg and 2 kg at points with coordinates (2,4), (0,3) and (5,1) relative to axes Ox and Oy.

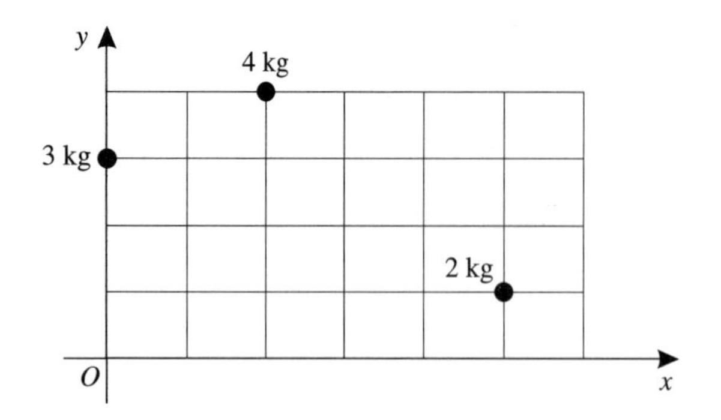

The first step is to find the total mass of the system.

The total mass of the system is $(4 + 3 + 2)$ kg $= 9$ kg.

Now form a table using the masses of the particles and their coordinates. The final column will contain the total mass and the unknown coordinates $(\bar{x}, \bar{y})$ of the centre of mass. Separate this column from the others with a double line.

	Separate masses			Total mass
Mass (kg)	4	3	2	9
x-coordinate	2	0	5	$\bar{x}$
y-coordinate	4	3	1	$\bar{y}$

Using the x-coordinates gives:

$$\Sigma m_i x_i = \Sigma m_i \bar{x}$$

$$4 \times 2 + 3 \times 0 + 2 \times 5 = 9\bar{x}$$

$$9\bar{x} = 18$$

$$\bar{x} = 2$$

Using the y-coordinates gives:

$$\Sigma m_i y_i = \Sigma m_i \bar{y}$$

$$4 \times 4 + 3 \times 3 + 2 \times 1 = 9\bar{y}$$

$$9\bar{y} = 27$$

$$\bar{y} = 3$$

The coordinates of the centre of mass are (2,3).

Example 3

Particles of masses 1 kg, 2 kg, 3 kg and 4 kg are attached to the corners of a light rectangular plate $ABCD$. Given that $AB = 10$ cm and $AD = 5$ cm calculate the distance of the centre of mass of the system (a) from AB (b) from AD.

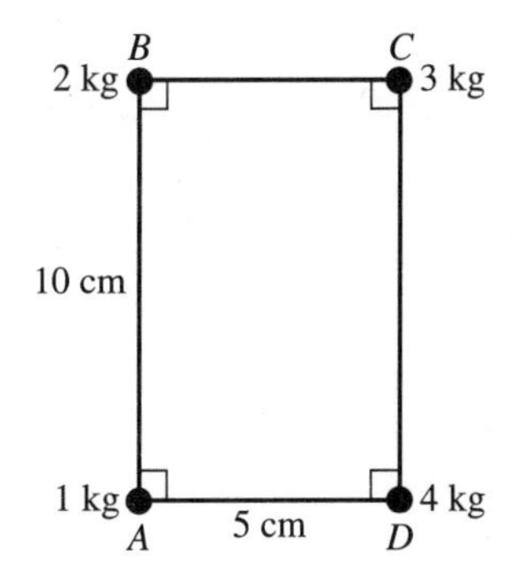

In this question no axes are given, but you need to find the distances of the centre of mass from AB and AD.

These lines are perpendicular to each other and are the best choice of axes.

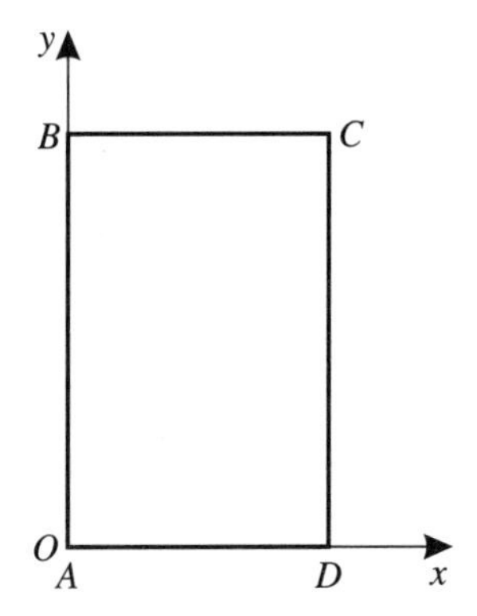

The total mass of the system is $(1 + 2 + 3 + 4)$ kg $= 10$ kg.

Let the centre of mass be at G where G is $\bar{x}$ cm from AB and $\bar{y}$ cm from AD.

	Separate masses				Total mass
Mass (kg)	1	2	3	4	10
Distance from AB (cm)	0	0	5	5	$\bar{x}$
Distance from AD (cm)	0	10	10	0	$\bar{y}$

Using the x-coordinates gives:

$$\Sigma m_i x_i = \Sigma m_i \bar{x}$$

$$1 \times 0 + 2 \times 0 + 3 \times 5 + 4 \times 5 = 10\bar{x}$$

$$10\bar{x} = 35$$

$$\bar{x} = 3.5$$

The centre of mass is 3.5 cm from AB.

Using the y-coordinates gives:

$$\Sigma m_i y_i = \bar{y}$$

$$1 \times 0 + 2 \times 10 + 3 \times 10 + 4 \times 0 = 10\bar{y}$$

$$10\bar{y} = 50$$

$$\bar{y} = 5$$

The centre of mass is 5 cm from AD.

The above method can also be used if the position of the centre of mass of the system is given and either the position or the mass of one of the particles needs to be determined.

Example 4

Particles A, B and C of mass 2 kg, 4 kg and 4 kg are situated at points with coordinates (3,0), (1,3) and (0,1) respectively. A fourth particle D of mass 2 kg is placed at the point with coordinates (x, y). Given that the centre of mass of the resulting system is at (1,2) determine the values of x and y.

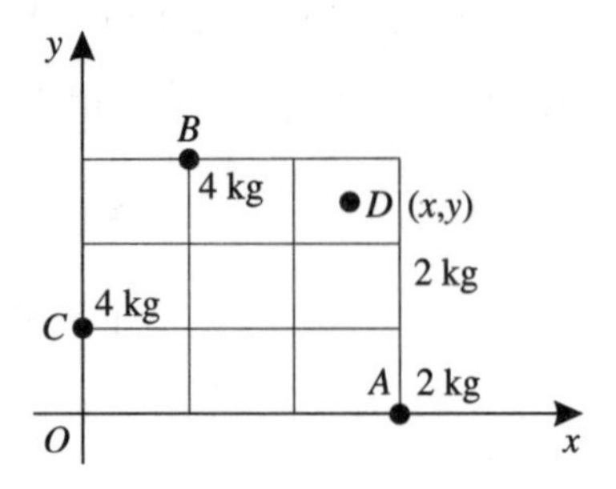

	Separate masses				Total mass
Mass (kg)	2	4	4	2	12
x-coordinate	3	1	0	x	1
y-coordinate	0	3	1	y	2

Using the x-coordinates gives:

$$\Sigma m_i x_i = \Sigma m_i \bar{x}$$
$$2 \times 3 + 4 \times 1 + 4 \times 0 + 2x = 12 \times 1$$
$$2x = 2$$
$$x = 1$$

Using the y-coordinates gives:

$$\Sigma m_i y_i = \Sigma m_i \bar{y}$$
$$2 \times 0 + 4 \times 3 + 4 \times 1 + 2y = 12 \times 2$$
$$2y = 8$$
$$y = 4$$

So x is 1 and y is 4.

If any of the particles are situated at points which have negative values for either or both coordinates the method used is the same but the signs of the coordinates must be included.

Example 5

Particles A, B, C and D of masses 1 kg, 2 kg, 1 kg and m kg respectively are situated at points with coordinates (5,–1), (–1,2), (6,4) and (1,0) respectively. Given that the centre of mass of the system is at (2, $\bar{y}$), determine the values of m and $\bar{y}$.

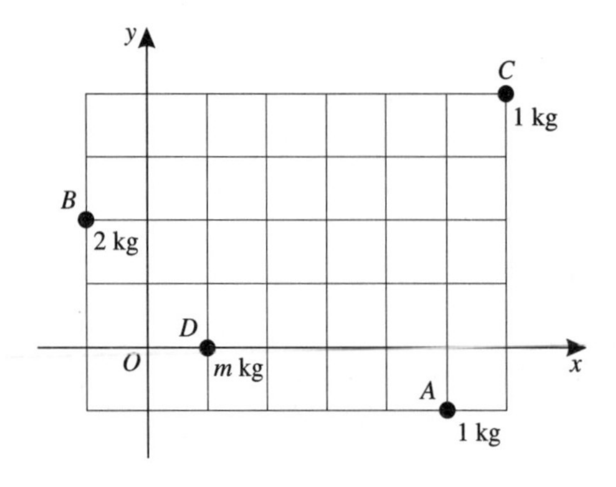

	Separate masses				Total mass
Mass (kg)	1	2	1	m	$4+m$
x-coordinate	5	−1	6	1	2
y-coordinate	−1	2	4	0	$\bar{y}$

Using the x-coordinates gives:

$$\Sigma m_i x_i = \Sigma m_i \bar{x}$$
$$1 \times 5 + 2 \times (-1) + 1 \times 6 + m \times 1 = (4+m)2$$
$$5 - 2 + 6 + m = 8 + 2m$$
$$1 = m$$

Using the y-coordinates gives:

$$\Sigma m_i y_i = \bar{y}$$
$$1 \times -1 + 2 \times 2 + 1 \times 4 + m \times 0 = (4+m)\bar{y}$$
$$-1 + 4 + 4 = 5\bar{y}$$

Substituting $m = 1$ gives:

$$7 = 5\bar{y}$$
$$\bar{y} = \tfrac{7}{5} = 1.4$$

So m is 1 and $\bar{y}$ is 1.4.

Exercise 2A

1 Three particles of masses 1 kg, 2 kg and 3 kg lie on the x-axis at points with coordinates (1,0), (3,0) and (6,0) respectively. Determine the coordinates of the centre of mass of the system.

2 Four particles of masses 4 kg, 3 kg, 2 kg and 5 kg lie on the y-axis at points with coordinates (0,2), (0,4), (0,5) and (0,6) respectively. Determine the coordinates of their centre of mass.

3 Three particles of masses 2 kg, 4 kg and 4 kg lie at the points with coordinates (1,1), (3,3) and (4,4) respectively. Determine the coordinates of their centre of mass.

4 A light rod AB of length 3 m has particles of masses 2 kg, 4 kg and 3 kg attached to it at the points A, B and C respectively where $AC = 1$ m. Determine the distance of the centre of mass from A.

5 Three particles of masses 3 kg, 5 kg and m kg lie on the x-axis at the points with coordinates (2,0), (4,0) and (5,0) respectively. The centre of mass of the system is at (4,0). Determine the value of m.

6 A light rod AB of length 2m has particles of masses 4 kg and 6 kg attached to it at the points A and C respectively where $AC = 0.5$ m. Determine the mass of the particle that must be attached at B if the centre of mass of the system is to be at the mid-point of the rod.

7 Three particles of masses 2 kg, 3 kg and 2 kg lie in the (x, y) plane at the points with coordinates (2,3), (3,6), and (−3,2) respectively. Determine the coordinates of the centre of mass of the system.

8 Four particles of equal mass lie in the (x, y) plane at the points with coordinates (3,5), (−1,2), (3,−4) and (−3, −2). Determine the coordinates of the centre of mass of the system.

9 A light rectangular plate $PQRS$ where $PQ = 4$ m and $PS = 3$ m has particles of masses 2 kg, 2 kg, 3 kg and 4 kg attached at P, Q, R and S respectively. Calculate the distance of the centre of mass of the system (a) from PQ (b) from PS.

10 A light rectangular plate $ABCD$ where $AB = 10$ cm and $AD = 9$ cm has particles of masses 3 kg, 2 kg and 4 kg attached to it at the points A, B and X respectively where X is the mid-point of CD. Calculate the distance of the centre of mass of the system from (a) AB (b) AD.

11 A light rectangular plate $ABCD$ where $AB = 5$ cm and $AD = 12$ cm has particles of masses 4 kg, 3 kg and 2 kg attached at A, B and C respectively. A fourth particle of mass 5 kg is to be attached to the plate. Given that the centre of mass of the resulting system is to be at the centre of the plate, determine the distance at which this particle must be attached (a) from AB (b) from AD.

12 Particles of masses 4 kg, 3 kg and 5 kg are attached to the corners A, B and C of a light rectangular plate $ABCD$ where $AB = 2$ m and $AD = 4$ m. Calculate the mass that must be attached at D if the centre of mass of the system is to be at a distance of 2 m from AB. Using this mass, calculate the distance of the centre of mass from AD.

13 A rectangular framework is made by joining light rods AB, BC, CD and AD where $AB = CD = 1\,\text{m}$ and $BC = AD = 2\,\text{m}$. Particles of masses 2 kg, 4 kg, 4 kg and 5 kg are placed at the mid-points of AB, BC, CD and AD respectively. Calculate the distance of the centre of mass of the resulting system from (a) AB (b) AD.

2.2 Centre of mass of a uniform plane lamina

A uniform plane lamina has its mass distributed evenly throughout its area. Any uniform plane lamina with an axis of symmetry has its mass evenly distributed on either side of the axis. Therefore the centre of mass of that figure must lie on the axis of symmetry. For a figure with more than one axis of symmetry, it follows that the centre of mass is at the point of intersection of these axes.

- **The symmetries of a plane lamina can be used to determine its centre of mass.**

Standard results

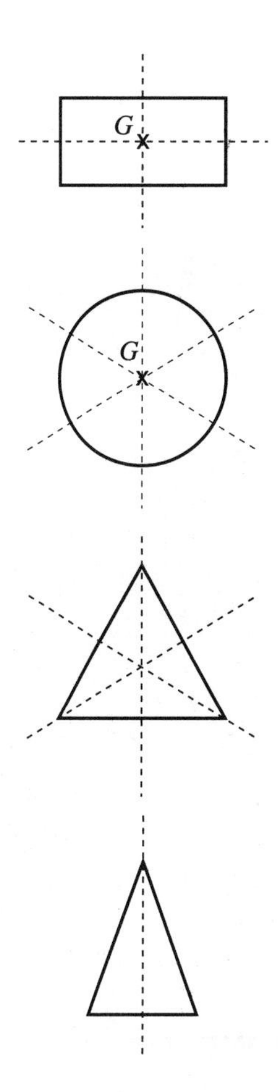

Uniform rectangular lamina
A uniform rectangular lamina has two axes of symmetry which are the two lines joining the mid-points of opposite sides. The centre of mass G is at their point of intersection.

Uniform circular disc
A uniform circular disc has every diameter as an axis of symmetry. The centre of the circle is therefore the centre of mass G of the disc.

Uniform triangular lamina
A uniform triangular lamina only has axes of symmetry if it is equilateral or isosceles.

A uniform equilateral triangle has three axes of symmetry which are called the **medians** of the triangle. These are the three lines which join the vertices of the triangle to the mid-points of the opposite sides. The centre of mass G is at their point of intersection.

A uniform isosceles triangle has only one axis of symmetry. Its centre of mass still lies at the point of intersection of the medians.

A scalene triangle has no axes of symmetry. It can be shown that the medians of any triangle intersect at a point which is the centre of mass G of the triangle. This point is $\frac{1}{3}$ distance along the median from the mid-point of that side. That is for triangle ABC below, $GD = \frac{1}{3}\,AD$.

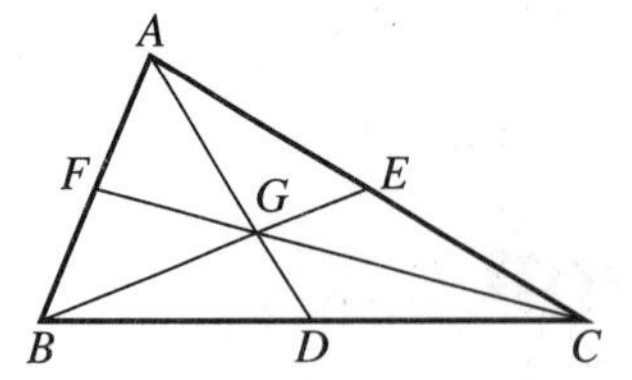

This result is explained in Book M3.

The centres of mass of the following uniform plane figures can be found by methods which will be explained in Book M3. In the M2 examination you may be required to use the results, which are quoted in the formula book provided by the examination board.

Uniform circular arc

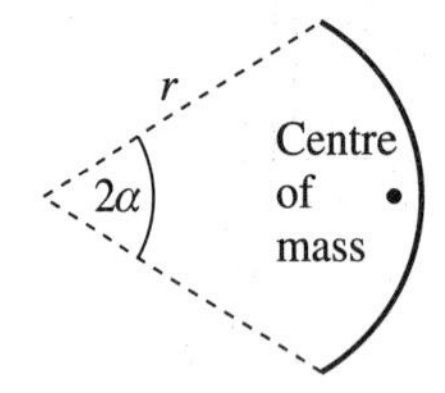

A uniform circular arc of radius r and angle at the centre 2α has its centre of mass on the axis of symmetry at a distance $\dfrac{r \sin \alpha}{\alpha}$ from the centre, *where* α *is measured in radians.*

Uniform sector

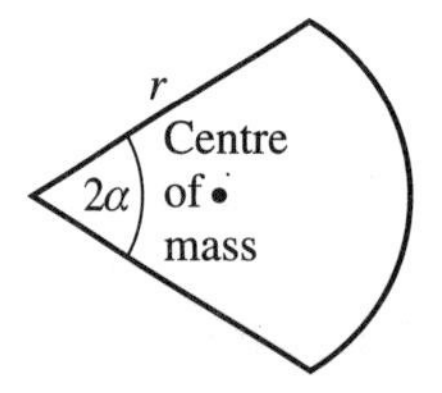

A uniform sector of a circle of radius r and angle at the centre 2α has its centre of mass on the axis of symmetry at a distance $\dfrac{2r \sin \alpha}{3\alpha}$ from the centre, *where* α *is measured in radians.*

Note that when the centre of mass of a semicircular wire or lamina is required it can be found by writing $\alpha = \dfrac{\pi}{2}$ in the appropriate formula.

Example 6

Calculate the coordinates of the centre of mass of triangle ABC with vertices at (0,0), (0,9) and (12,0).

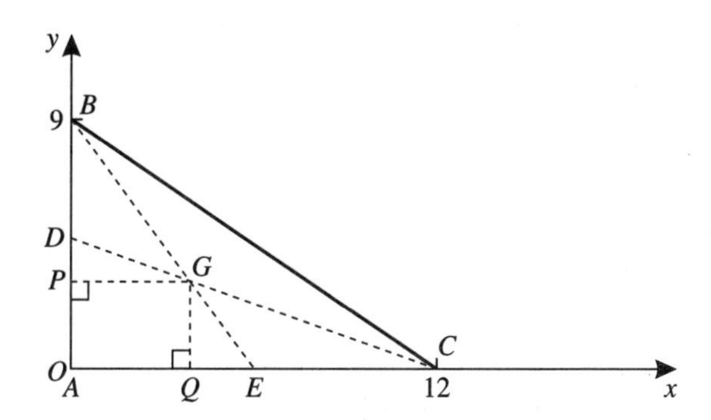

Let D be the mid-point of AB and E be the mid-point of AC.

Then the centre of mass is at G, where $CG = \frac{2}{3}\,CD$ and $BG = \frac{2}{3}\,BE$.

Because the triangles ADC and QGC are similar:

$$\frac{CG}{CD} = \frac{CQ}{AC}$$

And because: $\dfrac{CG}{CD} = \frac{2}{3}$

It follows that: $\dfrac{CQ}{AC} = \frac{2}{3}$

Or: $CQ = \frac{2}{3}AC$

Then $AQ = \frac{1}{3}\,AC$ and the x coordinate of G is 4.

Also because the triangle BPG and BAE are similar:

$$\frac{BP}{BA} = \frac{BG}{BE} = \frac{2}{3}$$

So: $\mathrm{BP} = \frac{2}{3}\,\mathrm{AB}$ and $\mathrm{AP} = \frac{1}{3}\,\mathrm{AB}$.

The y coordinate of G is 3.

The coordinates of G are (4,3).

In questions like the above, where the answer can be ‘seen’ from the geometry of the diagram it is not necessary to write down a lengthy calculation.

Composite plane figures

Combining two or more uniform plane figures, like those above, will produce a composite figure.

You can calculate the position of the centre of mass of a composite figure by using the same methods as before with the standard results given above.

If the composite figure has an axis of symmetry then the centre of mass will lie on that axis.

Example 7

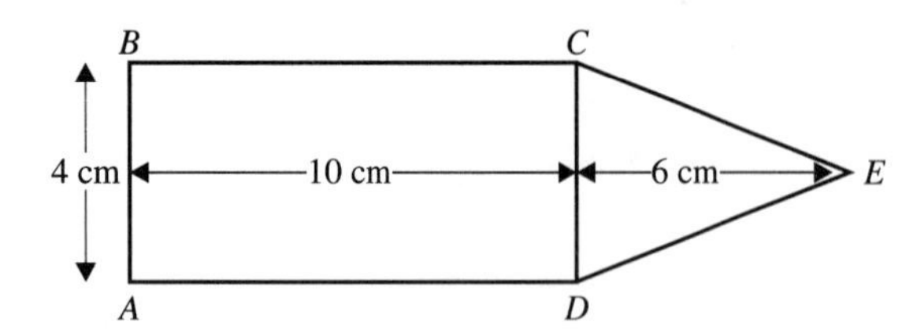

This diagram shows a uniform lamina formed by joining a rectangular lamina $ABCD$ and an isosceles triangular lamina CDE, both of mass per unit area m kg. Calculate the distance of the centre of mass of $ABCED$ (a) from AB (b) from AD.

Take the x- and y-axes to be in the directions of AD and AB respectively.

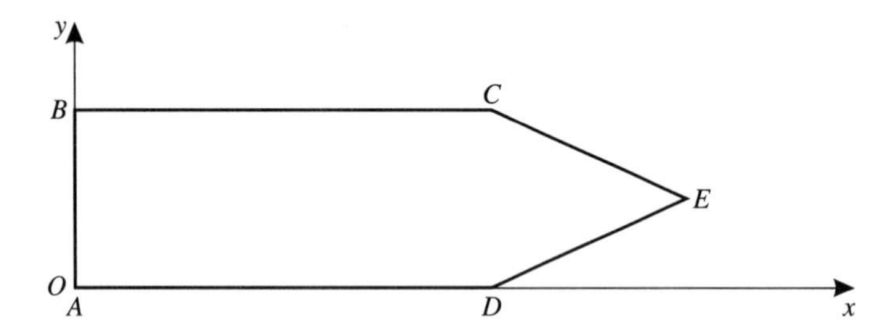

The area of $ABCD$ is:

$$(4 \times 10) = 40\,\text{cm}^2$$

And the mass of $ABCD = 40m$ kg.

Similarly the mass of CDE is:

$$(\tfrac{1}{2} \times 4 \times 6\text{m}) = 12m\,\text{kg}.$$

So the total mass of $ABCED$ is $(40m + 12m) = 52m$ kg.

Construct a table showing the masses and coordinates of the centres of masses of the separate parts and of the total mass.

The distance $\bar{x}$ is the distance of the centre of mass from AB and the distance $\bar{y}$ is the distance of the centre of mass from AD.

	Separate masses		Total mass
Mass (kg)	$40m$	$12m$	$52m$
Distance x (cm)	5	$(10 + 2) = 12$	$\bar{x}$
Distance y (cm)	2	2	$\bar{y}$

(a) Using the x-coordinates gives:

$$40\,m \times 5 + 12m \times 12 = 52\,m \times \bar{x}$$

$$\bar{x} = \frac{344}{52} = 6.62$$

The centre of mass is 6.62 cm from AB.

(b) Using the y-coordinates gives:

$$40m \times 2 + 12m \times 2 = 52\,m\bar{y}$$

$$\bar{y} = \frac{104}{52} = 2$$

The centre of mass is 2 cm from AD.

There is an alternative solution for (b).

If you see an axis of symmetry for the figure you can use this to obtain your solution but you must state that you are using symmetry.

In example 7, the lamina has an axis of symmetry which is the line joining the mid-point of AB to E. The centre of mass must lie on this line. This can be used to solve part (b) of the question.

In this case, using symmetry for (b), the centre of mass is 2 cm from AD.

Because m, the mass per unit area, was a factor for every term in both the equations in example 7, all the masses cancelled out. When calculating the position of the centre of mass of a uniform lamina you only need to use the ratio of the masses as given by the areas.

If you are required to find the position of the centre of mass of a lamina which has had part of its original area cut away, use the same method so long as the complete lamina is considered to be formed from the final lamina and the part cut away.

Example 8

An ear-ring is formed from a thin uniform square of metal $ABCD$ of side 2 cm by cutting away a square $PQRS$ of side 0.6 cm, as shown in the diagram.

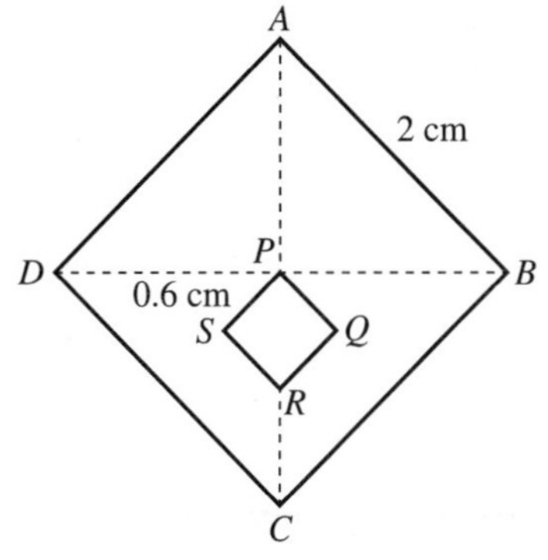

P lies on the diagonal BD of $ABCD$ and PR lies along the diagonal AC of $ABCD$.

Find the position of the centre of mass of the ear-ring.

Diagonal AC is an axis of symmetry of the ear-ring.

So the centre of mass lies on AC.

You need the distances from A of the centres of mass of the two squares $ABCD$ and $PQRS$.

Centre of mass of $ABCD$ is at P.

$$AP = \tfrac{1}{2} AC$$

By Pythagoras' Theorem:

$$AC = \sqrt{(2^2 + 2^2)} = \sqrt{8} \text{ cm}$$

Hence: $$AP = \frac{\sqrt{8}}{2} = 1.414$$

So the distance of the centre of mass of $ABCD$ from A is 1.414 cm.

Centre of mass of $PQRS$ is at the mid-point of PR.

$$PR = \sqrt{(0.6^2 + 0.6^2)}$$
$$= 0.8485$$

So the distance of the centre of mass of $PQRS$ from A is:

$$(1.414 + \tfrac{1}{2} \times 0.8485) \text{ cm}$$
$$= 1.838 \text{ cm}$$

Let the centre of mass of the ear-ring be distance $\bar{x}$ cm from A, measured along AC.

$$\text{Area of } PQRS = 0.36 \text{ cm}^2$$
$$\text{Area of ear-ring} = 4 - 0.36$$
$$= 3.64 \text{ cm}^2$$

	Separate masses		Total mass
Ratio of masses	0.36	3.64	4
Distance of C of M from A	1.838	$\bar{x}$	1.414

Taking moments gives:

$$0.36 \times 1.838 + 3.64\bar{x} = 4 \times 1.414$$

$$\bar{x} = \frac{4 \times 1.414 - 0.36 \times 1.838}{3.64}$$

$$\bar{x} = 1.37.$$

The centre of mass is on AC at a distance 1.37 cm from A.

Centre of mass of a uniform rod and of a framework

The centre of mass of a uniform rod is at the mid-point of the rod.

A **framework** is a body which is formed by joining a number of rods or wires. The position of the centre of mass of a framework can be calculated by using the positions of the centres of mass of the separate rods or wires.

Example 9

A uniform wire of length 30 cm is bent to form a triangle ABC where AB has length 5 cm and AC has length 12 cm. Calculate the distance of the centre of mass (a) from AB (b) from AC.

The perimeter of the triangle is 30 cm and two sides are 5 cm and 12 cm. So BC is 13 cm and the triangle is right-angled at A.

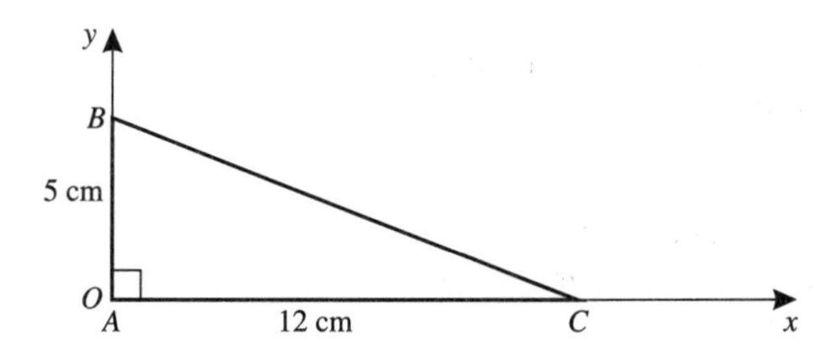

Take axes along AC and AB as shown in the diagram. Since the wire is uniform the mass of each side of the triangle will be proportional to its length. Each side of the triangle is a uniform rod and so the centre of mass of each side is at its mid-point.

Let the centre of mass of the triangle have coordinates $(\bar{x}, \bar{y})$ where $\bar{x}$ is the distance of the centre of mass from AB and $\bar{y}$ is the distance of the centre of mass from AC.

	Separate masses			Total mass
Ratio of masses	5	12	13	30
x-coordinate	0	6	6	$\bar{x}$
y-coordinate	2.5	0	2.5	$\bar{y}$

Using the x-coordinates gives:

$$5 \times 0 + 12 \times 6 + 13 \times 6 = 30\bar{x}$$

$$72 + 78 = 30\bar{x}$$

$$\bar{x} = \frac{150}{30} = 5$$

The centre of mass is 5 cm from AB.

Using the y-coordinates gives:

$$5 \times 2.5 + 12 \times 0 + 13 \times 2.5 = 30\bar{y}$$

$$\bar{y} = \frac{45}{30} = 1.5$$

The centre of mass is 1.5 cm from AC.

Example 10

An ear-ring is formed by bending a uniform wire until it forms the outline of a sector of a circle. The two straight sides AB and AC have length 2 cm and the angle between the sides is 40°. Find the distance of the centre of mass of the ear-ring from A.

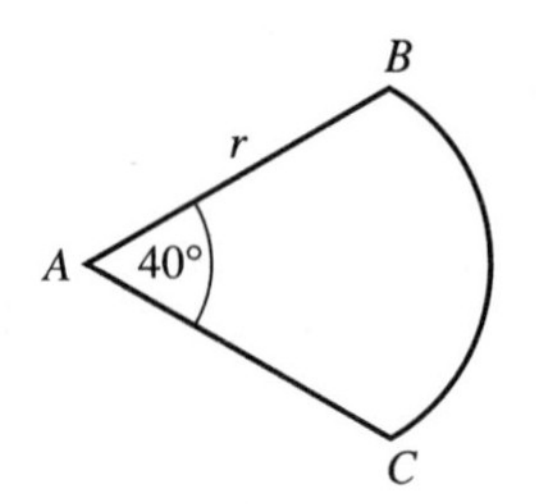

Take the axis of symmetry to be the x-axis. The centre of mass lies on the axis of symmetry.

Since the wire is uniform the mass of each of the three parts, AB, BC and AC, of the ear-ring is proportional to its length, so the length of the arc BC is required.

The angle α (20°) must be in radians when calculating the distance of the centre of mass of the arc BC from A. It is therefore preferable to work in radians throughout the question.

We have: $$20° = 20 \times \frac{\pi}{180} = \frac{\pi}{9} \text{ radians}$$

The radius of the circle is AB which measures 2 cm, so:

$$\text{arc length } BC = 2 \times 2 \times \frac{\pi}{9} = \frac{4\pi}{9} \text{ cm}$$

Using the standard formula (from the formula book) gives:

$$\begin{array}{c}\text{distance of centre of mass}\\ \text{of arc } BC \text{ from } A\end{array} = \frac{r \sin \alpha}{\alpha} = \frac{2 \sin \frac{\pi}{9}}{\frac{\pi}{9}} = \frac{18 \sin \frac{\pi}{9}}{\pi} \text{ cm}$$

Let the centre of mass of the ear-ring be distance $\bar{x}$ cm from A.

	Separate masses			Total mass
Ratio of masses	2	2	$\frac{4\pi}{9}$	$4 + \frac{4\pi}{9}$
x-coordinate	$1 \times \cos\frac{\pi}{9}$	$1 \times \cos\frac{\pi}{9}$	$\frac{18 \sin \frac{\pi}{9}}{\pi}$	$\bar{x}$

Taking moments gives:

$$2 \times \cos\frac{\pi}{9} + 2 \times \cos\frac{\pi}{9} + \frac{4\pi}{9} \times \frac{18\sin\frac{\pi}{9}}{\pi} = \left(4 + \frac{4\pi}{9}\right)\bar{x}$$

$$4\cos\frac{\pi}{9} + 8\sin\frac{\pi}{9} = \left(4 + \frac{4\pi}{9}\right)\bar{x}$$

$$\bar{x} = \frac{\left(4\cos\frac{\pi}{9} + 8\sin\frac{\pi}{9}\right)}{4 + \frac{4\pi}{9}}$$

$$\bar{x} = 1.20$$

The centre of mass of the ear-ring is 1.20 cm from A.

Exercise 2B

1 For each of the following uniform laminae write down the coordinates of the centre of mass :

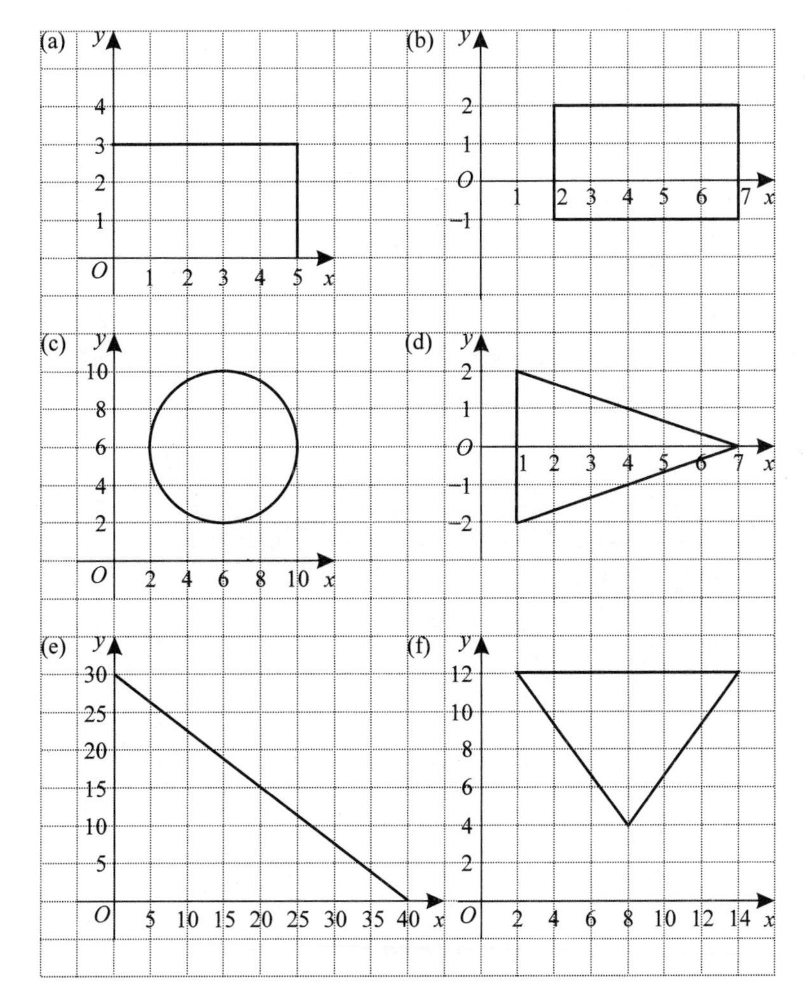

2 For each of the following uniform laminae calculate the coordinates of the centre of mass.

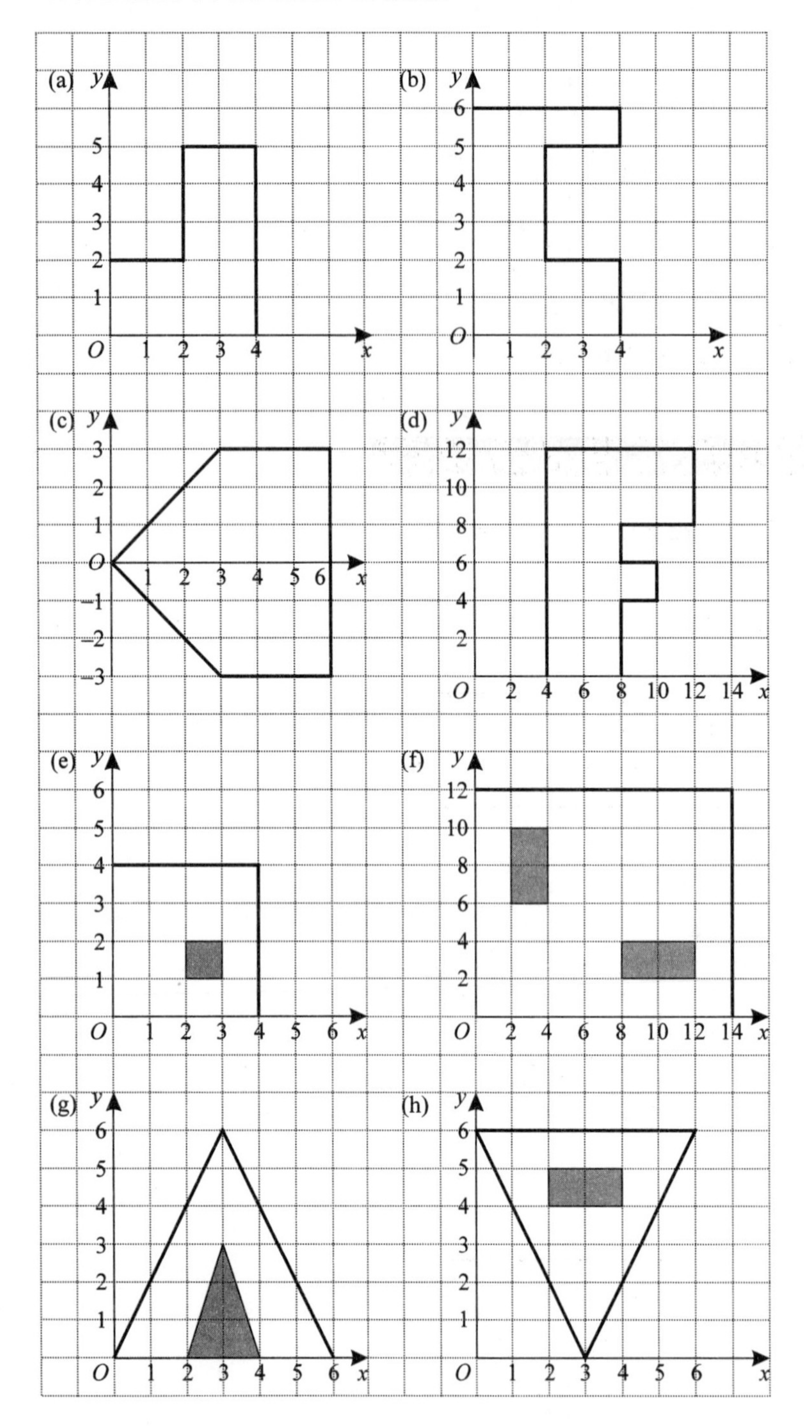

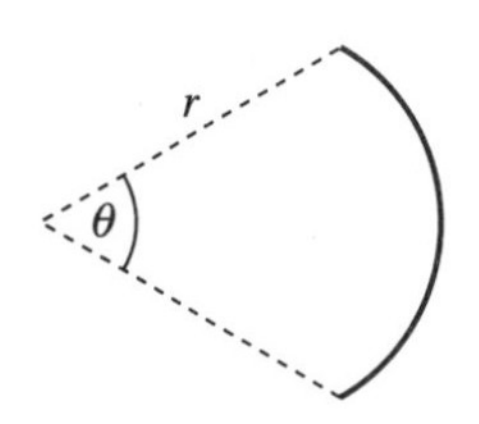

3 An arc of a circle has radius r cm and the angle at the centre is θ. Find the distance of the centre of mass of the arc from the centre of the circle when:

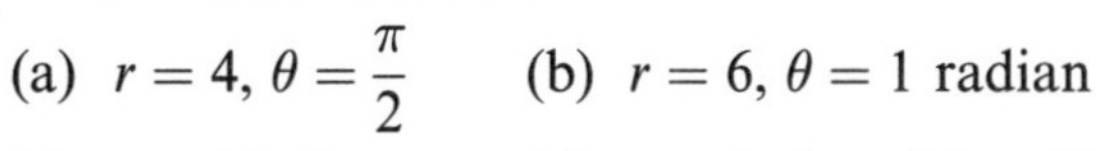

(a) $r = 4$, $\theta = \dfrac{\pi}{2}$ (b) $r = 6$, $\theta = 1$ radian

(c) $r = 12$, $\theta = \pi$ (d) $r = 5$, $\theta = 60°$ (e) $r = 8$, $\theta = 120°$

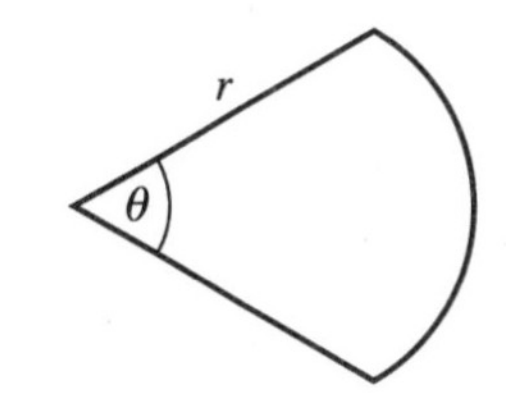

4 A sector of a circle has radius r cm and the angle at the centre is θ. Find the distance of the centre of mass of the sector from the centre of the circle when:

(a) $r = 9$, $\theta = \pi$ radians (b) $r = 6$, $\theta = \dfrac{\pi}{3}$ radians

(c) $r = 8$, $\theta = 0.5$ radians (d) $r = 12$, $\theta = 90°$

(e) $r = 10$, $\theta = 45°$

5 A uniform wire of length 6 m is bent to form a triangle ABC where AB is 1.5 m and BC is 2.5 m. Calculate the distance of the centre of mass of the triangle (a) from AB (b) from AC.

6 Three uniform rods, all of the same mass per unit length, are joined to form a triangle ABC where AB is 7 cm, BC is 24 cm and AC is 25 cm. Calculate the distance of the centre of mass of the triangle (a) from AB (b) from BC.

7 Three uniform rods AB, BC and AC are of length 3 m, 4 m and 5 m and have masses 5 kg, 4 kg and 3 kg respectively. They are joined to make a framework. Calculate the distance of the centre of mass of the framework (a) from AB (b) from BC.

8 A uniform circular lamina of radius 5 units has its centre at (2,0). Two circles are cut away from this lamina, one having centre (0,0) and radius 2 units and the other having centre (3,2) and radius 1 unit. Calculate the coordinates of the centre of mass of the resulting lamina.

9 A uniform square metal plate of side 10 cm has a square of side 4 cm cut away from one corner. Find the position of the centre of mass of the remaining plate.

10 The diagram shows a thin uniform square of metal which has a corner folded over. Calculate the distance of the centre of mass of $ABCEF$ from the corner A.

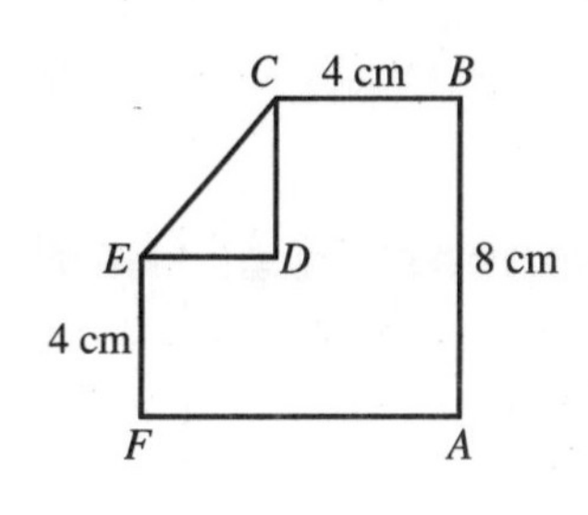

11 A semicircular lamina of radius 6 cm has a circular hole of radius 1.5 cm cut in it. The centre of the hole is on the axis of symmetry of the semicircle at a distance 2 cm from the straight edge of the lamina. Find the distance of the centre of mass of the lamina from its straight edge.

12 The diagram shows a uniform rectangular lamina with a semicircle cut away from one edge. The resulting shape is symmetrical. The rectangle measures 20 cm by 10 cm and the semicircle has radius 8 cm. Find the position of the centre of mass from the centre of the semicircle.

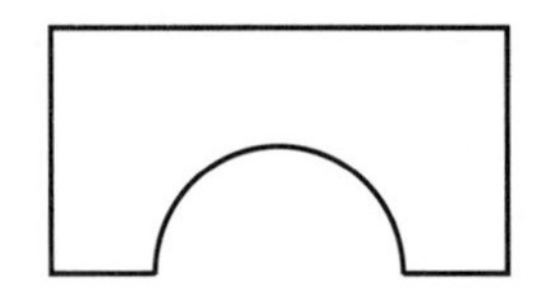

13 The diagram shows a pendant in the shape of a sector of a circle with centre A. The radius is 4 cm and the angle at A is 0.4 radians. Three small holes of radius 0.1 cm, 0.2 cm and 0.3 cm are cut away. The diameters of the holes lie along the axis of symmetry and their centres are 1 cm, 2 cm and 3 cm respectively from A. The pendant can be modelled as a uniform lamina. Find the distance of the centre of mass of the pendant from A.

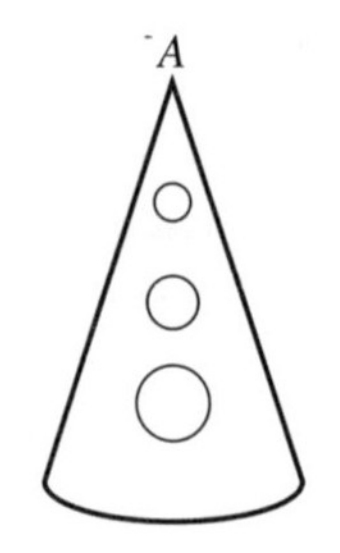

14 A uniform wire is bent to form the outline of a sector of a circle, with the wire being doubled along the arc only. Given that the straight sides measure 0.5 m and the angle between them is 40°, calculate the distance of the centre of mass of the framework from the centre of the circle.

15 The diagram shows a uniform semicircular lamina of mass M. A is the mid-point of the diameter and B is on the circumference at the other end of the axis of symmetry. A particle of mass m is attached to the lamina at B. The centre of mass of the loaded lamina is at the mid-point of AB. Find, in terms of π, the ratio $M : m$.

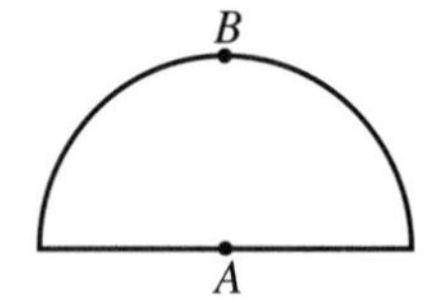

2.3 Equilibrium of a plane lamina

A lamina can be in equilibrium in a vertical plane when it is suspended from a point or is balanced on an edge. For either of these situations you need to know the position of the centre of mass of the lamina before you can proceed but the situations themselves are very different and so they will be considered separately.

Suspension of the lamina from a fixed point

If a lamina is suspended by a string and allowed to hang freely there are only two forces acting on it – its weight and the tension in the string. Consider a triangle ABC suspended from vertex A.

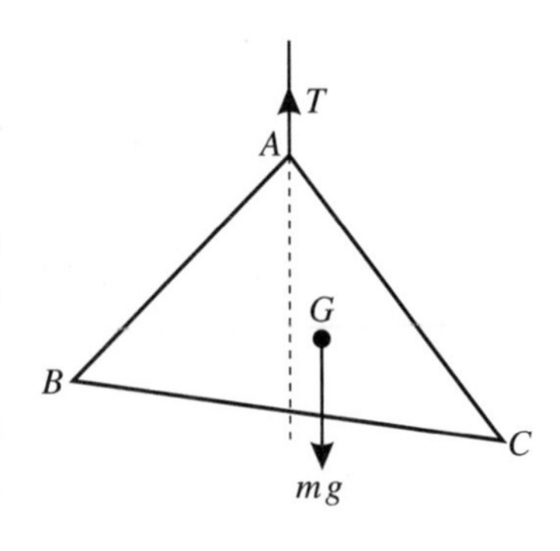

ABC could not hang as shown in the diagram because its weight would have a clockwise moment about A and the tension would have a zero moment about the same point. Hence the total moment cannot be zero as needed for equilibrium. The lamina must be in the position where the line of action of the weight passes through A so that there is no resultant moment about A.

A lamina may be suspended from a fixed point in other ways. It could be suspended by means of a horizontal nail through a point of the lamina, which could then be free to rotate about the fixed horizontal axis formed by the nail. In this chapter only cases of static equilibrium are being considered so this alternative means of suspension is basically no different from the suspension by means of a string.

Consider once more the triangle ABC, this time suspended by means of a nail through a point P of the lamina and free to rotate about the resulting horizontal axis. The lamina again has only two forces acting on it – its weight and the reaction from the nail. As the lamina can rotate it will move to its position of equilibrium with the two forces acting in the same vertical line.

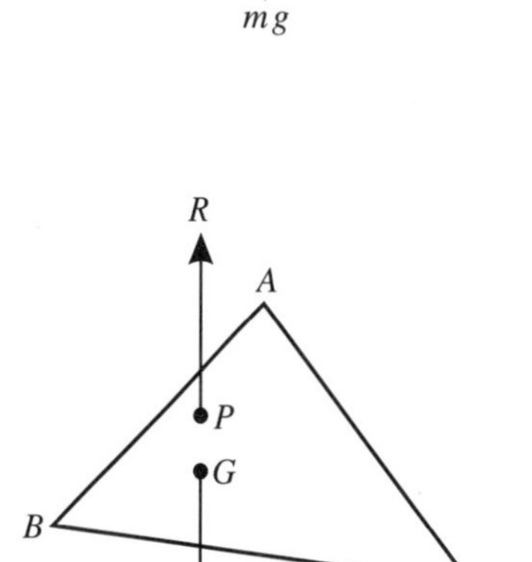

- **For a suspended lamina to be in equilibrium its centre of mass, G, must be vertically below the point of suspension.**

Example 11

The lamina shown in the diagram is suspended freely from A. Find the angle between AB and the vertical.

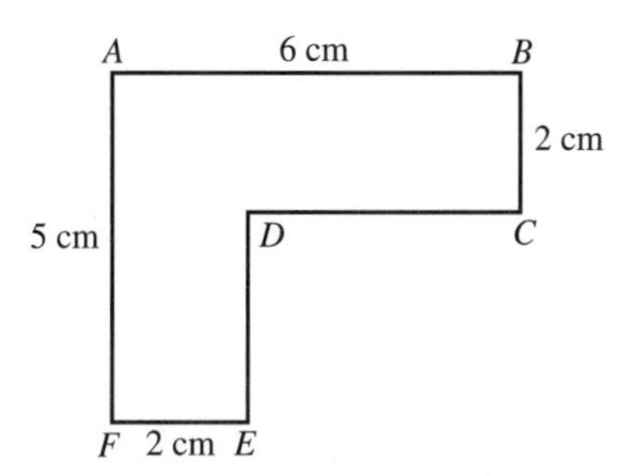

First find the position of the lamina's centre of mass, G. Consider the lamina to be formed of two rectangles $ABCP$ and $PDEF$.

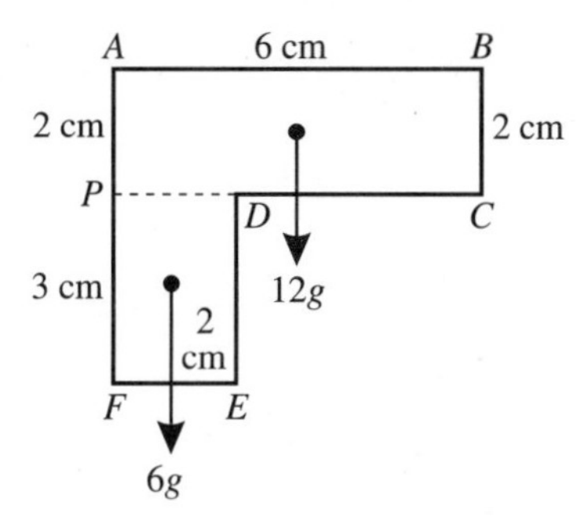

Let the centre of mass have coordinates $(\bar{x}, \bar{y})$ where $\bar{x}$ is the distance of the centre of mass from AF and $\bar{y}$ is the distance from AB.

	Separate masses		Total mass
Ratio of masses	12	6	18
Distance x (cm)	3	1	$\bar{x}$
Distance y (cm)	1	3.5	$\bar{y}$

Using the x-coordinates gives: $12 \times 3 + 6 \times 1 = 18\bar{x}$

$$\bar{x} = \frac{42}{18} = 2.333$$

Using the y-coordinates gives: $12 \times 1 + 6 \times 3.5 = 18\bar{y}$

$$\bar{y} = \frac{33}{18} = 1.833$$

To find the angle between AB and the vertical when the lamina is suspended, it is not necessary to re-draw the lamina. Simply add G to your original diagram, join AG and label this line as being vertical.

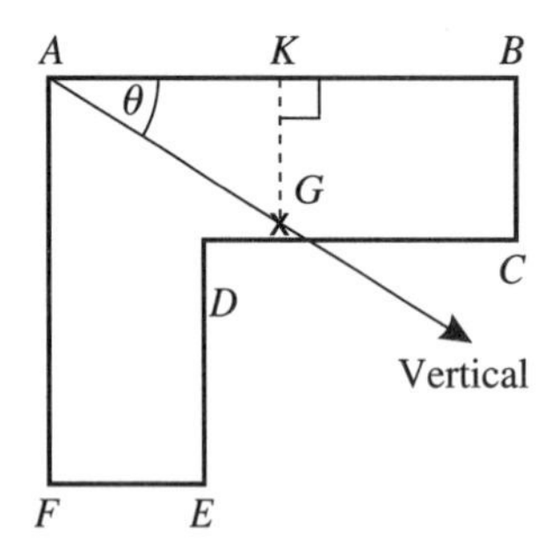

Mark the angle required as θ on your diagram and add the perpendicular line GK from G to AB to make a right-angled triangle. The lengths of AK and GK are the distances of G from AF and AB respectively.

So: $$\tan\theta = \frac{GK}{AK} = \frac{1.833}{2.333}$$

$$\theta = 38.16°$$

AB is inclined at 38.2° to the vertical.

Equilibrium of a lamina on an inclined plane

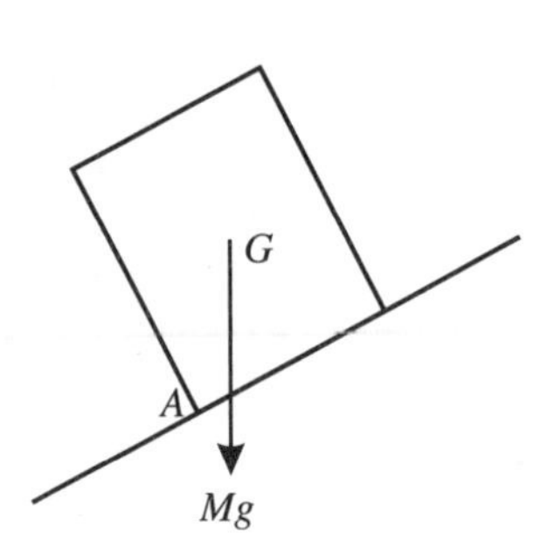

If a lamina is balanced on an inclined plane, the line of action of the weight of the lamina falls within the side of the lamina in contact with the plane. The weight will have a clockwise moment about the point A but the plane prevents the lamina from turning.

The angle of the plane could be increased, so that the situation is as shown opposite.

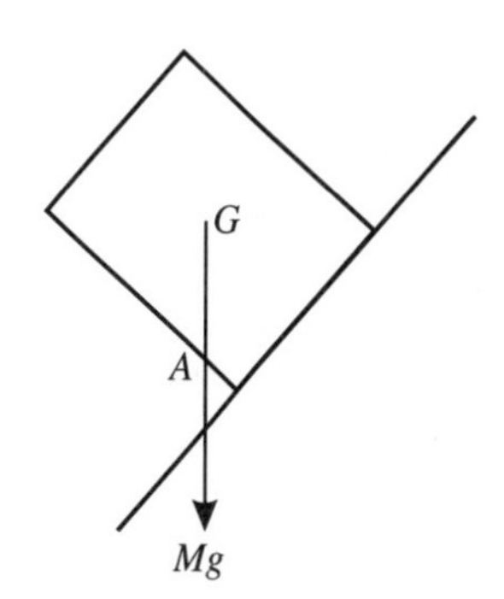

The weight now has an anticlockwise moment about A and the plane is no longer preventing the lamina from turning. In this case equilibrium is not possible.

- **For a lamina on an inclined plane to be in equilibrium the line of action of the weight of the lamina must fall within the side of the lamina which is in contact with the plane.**

Example 12

$ABCD$ is a rectangular lamina with $AB = 12\,\text{cm}$ and $AD = 5\,\text{cm}$. The lamina rests in equilibrium on an inclined plane with AD in contact with the plane. Find the maximum possible angle of inclination of the plane.

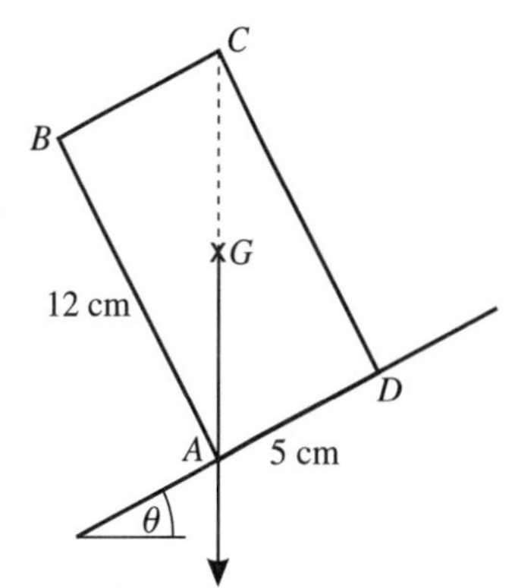

Let the plane be inclined at angle θ.

Let G be the centre of mass of the lamina.

θ will be maximum when AG is vertical, since increasing the angle of the plane any more will cause the line of action of the weight to fall outside the line of contact between the lamina and the plane.

From the geometry of the diagram $\angle BAG = \theta$.

G is at the mid-point of the diagonals of $ABCD$.

So A, G and C are collinear.

Therefore: $\tan \angle BAG = \frac{5}{12}$

$\angle BAG = 22.6°$

So: $\theta = 22.6°$

The maximum angle of inclination of the plane is 22.6°.

Exercise 2C

1 A uniform rectangular lamina $ABCD$ has $AB = 5\,\text{cm}$ and $AD = 8\,\text{cm}$. It is free to rotate about a fixed horizontal axis at A and hangs freely under gravity. Calculate the angle between AB and the downwards vertical.

2 A uniform triangular lamina ABC where $AB = 1\,\text{m}$ and $BC = 2\,\text{m}$ is right-angled at B. It is suspended from C and hangs freely under gravity. Calculate the angle between BC and the downwards vertical.

3 The framework described in Question 7 of Exercise 2B is suspended from B and hangs freely under gravity. Calculate the angle between AB and the vertical.

4 The wire described in Question 5 of Exercise 2B is hanging freely under gravity. Calculate the angle between AC and the vertical (a) when the wire is suspended from A (b) when the wire is suspended from C.

5 The ear-ring shown in the diagram is made from a uniform sheet of plastic. It consists of a semicircle centre O of radius 2 cm from which a concentric semicircle of radius 1 cm has been cut away. Given that the centre of mass of a uniform semicircular lamina of radius r is on the axis of symmetry at a distance $\frac{4r}{3\pi}$ from the straight edge, model the ear-ring as a uniform lamina and calculate the distance of its centre of mass from O. The ear-ring is suspended from point A and hangs freely under gravity. Calculate the angle between AB and the vertical, giving your answer to the nearest degree.

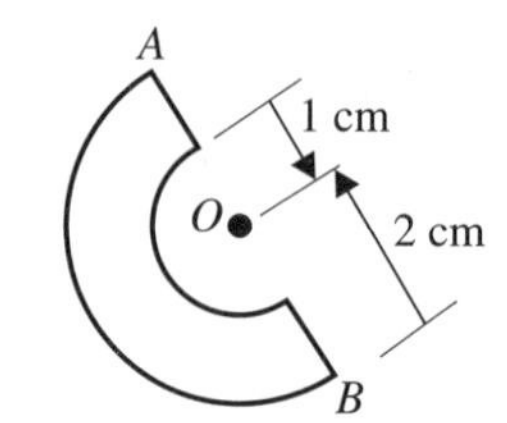

6 An engineer's square is in the shape of two uniform rectangular pieces of metal joined as shown in the diagram. The mass of $BCDE$ is three times the mass of $ABFG$. By modelling each rectangle as a uniform lamina calculate the distance of the centre of mass of the square (a) from AC (b) from CD.

The square is placed on a peg at F. By modelling the square as a lamina freely suspended from F, calculate the angle between FE and the vertical when the square hangs in equilibrium. State two assumptions you have made when using this model to calculate the angle.

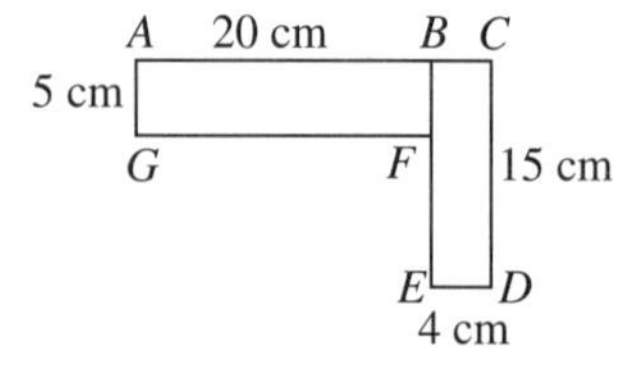

7 A uniform rectangular lamina $ABCD$ where $AB = 10$ cm and $BC = 5$ cm is placed on a rough inclined plane as shown in the diagram.

(a) Determine whether the rectangle can remain in equilibrium when the plane is inclined at an angle of 25°.

(b) Calculate the maximum possible angle of inclination of the plane if the rectangle is to be balanced as shown.

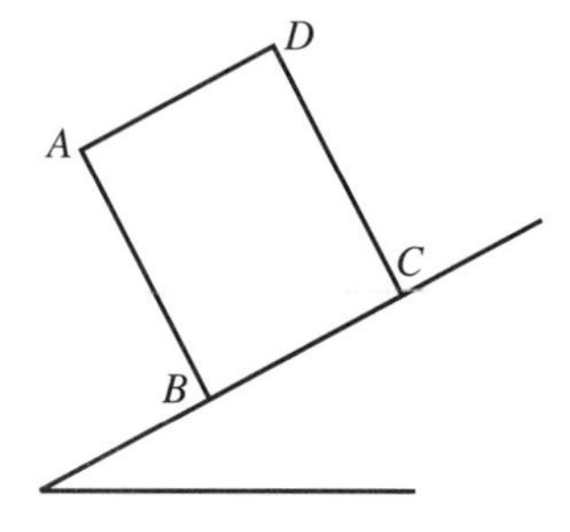

8 The metal plate described in Question 9 of Exercise 2B is placed on a rough inclined plane as shown in the diagram. The plane is inclined to the horizontal at an angle θ. Determine whether equilibrium is possible (a) when $\theta = 10°$ (b) when $\theta = 25°$.
Calculate the maximum possible angle of inclination of the plane for the lamina to be in equilibrium.

9 A uniform triangular sheet of metal ABC where $AB = 6$ cm and $AC = 12$ cm is right-angled at A. The triangle is placed on a plane inclined to the horizontal at angle θ as shown in the diagram.
By modelling the triangle as a uniform lamina determine the maximum value of θ for the triangle to remain in equilibrium in this position. As the angle of inclination of the plane is increased from zero, the equilibrium could be broken in a different way from tilting. State the form of motion that could have taken place. State also the assumption you made about your original model so that this form of motion did not take place.

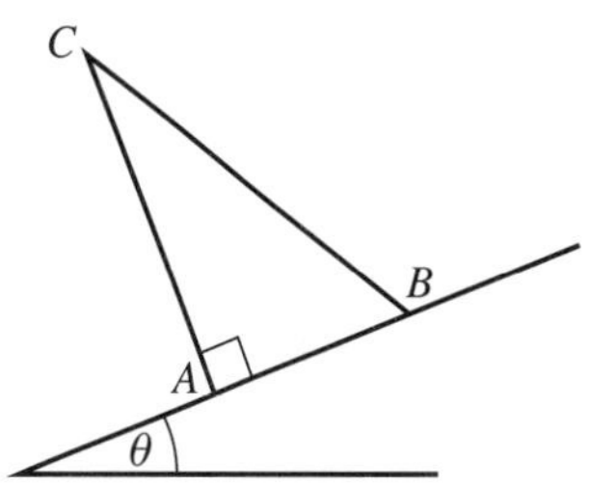

10 The engineer's square described in Question 6 of this exercise is placed on a rough plane inclined at angle θ to the horizontal as shown in the diagram.

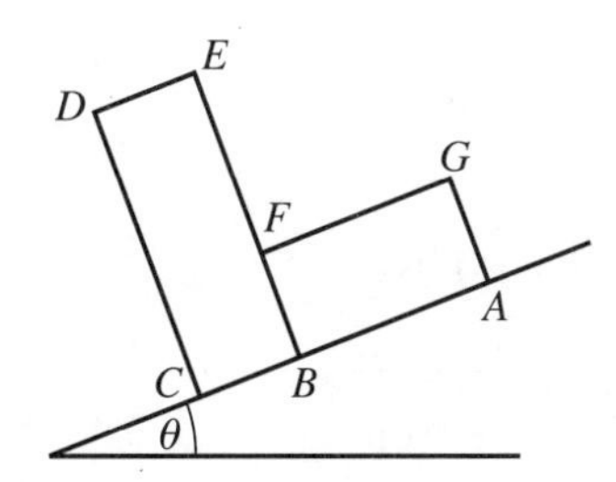

Find the maximum value of θ if the square is to remain in equilibrium.
The square is now placed on the inclined plane as shown in the second diagram.
Determine whether equilibrium is possible (a) when $\theta = 10°$ (b) when $\theta = 30°$.

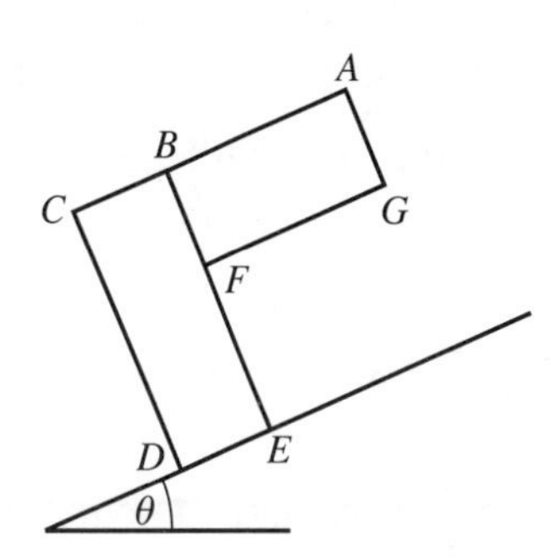

11 A plane figure is made from a uniform sheet of metal. The shape hangs in equilibrium from the point A. The centre of mass of the shape is G.

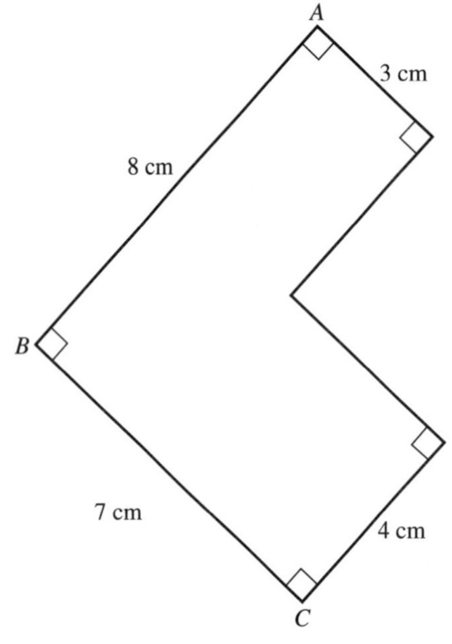

(a) Find the distance, in cm, of G from AB.

(b) Find the distance, in cm, of G from BC.

(c) Find, to the nearest degree, the angle made by AB with the downward vertical.

12 A uniform semicircular lamina is suspended from an end point of its diameter. Assuming that it hangs freely under gravity, calculate in radians the angle between the diameter and the vertical.

13 The lamina described in Question 10 of Exercise 2B is suspended from the corner B and hangs freely under gravity. Calculate in degrees the angle between AB and the vertical.

14 The diagram shows a uniform lamina. $ABCD$ is a rectangle which measures 12 cm by x cm. A semicircle of radius 6 cm is attached to the rectangle with the diameter of the semicircle coinciding with AB as shown. The centre of mass of the composite lamina lies on AB.

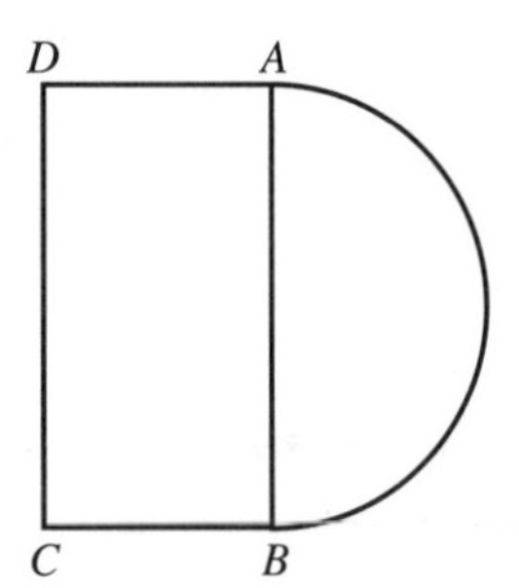

(a) Calculate the value of x.

The lamina is suspended from C and hangs freely under gravity.

(b) Calculate in degrees the angle between CD and the vertical.

SUMMARY OF KEY POINTS

1 The centre of mass of a lamina is the point at which the weight acts.

The weight of a uniform lamina is evenly distributed.

The centre of mass for a lamina or discrete mass distribution must lie on any axis of symmetry.

The centre of mass of a set of n masses $m_1, m_2, \ldots m_n$ at points with coordinates (x_1, y_1), (x_2, y_2), $\ldots$ (x_n, y_n) has coordinates $(\bar{x}, \bar{y})$ where:

$$\bar{x} = \frac{\Sigma m_i x_i}{\Sigma m_i}$$

and:
$$\bar{y} = \frac{\Sigma m_i y_i}{\Sigma m_i}$$

2 Standard results:

Body	**Centre of mass**
Uniform rod	mid-point of rod
Uniform rectangular lamina	point of intersection of lines joining mid-points of opposite sides
Uniform circular disc	centre of circle
Uniform triangular lamina	point of intersection of medians – in other words $\frac{2}{3}$ distance from any vertex to the mid-point of the opposite side
Circular arc, radius r, angle at centre 2α	$\frac{r \sin \alpha}{\alpha}$ from centre
Sector of circle, radius r, angle at centre 2α	$\frac{2r \sin \alpha}{3\alpha}$ from centre

3 For a lamina which is freely suspended and hangs in equilibrium, the centre of mass will be vertically below the point of suspension.

For a lamina which is balanced on an inclined plane, the line of action of the weight must fall within the side of the lamina that is in contact with the plane.

Work, energy and power

3

Work, energy and power are words most people are familiar with and use frequently without fully understanding their meaning. All three quantities result from the application of forces to objects and so have mechanical definitions as shall be seen in this chapter.

3.1 Work

If the point of application of a force F newtons moves through a distance s metres in the direction of the force then the **work done by the force** is given by:

■ $$\textit{Work} = F \times s$$

For a force in newtons and a distance in metres, the work done is measured in joules, abbreviated J.

Example 1

A packing case is pushed 5 m across a horizontal floor by a horizontal force of magnitude 30 N. Find the work done by the force.

30 N

Using: $\textit{Work} = F \times s$

Gives: $\textit{Work} = 30 \times 5 = 150$

The work done by the force is 150 J.

Example 2

The figure shows a box which is pulled at a constant speed across a horizontal surface by a horizontal rope. When the box has moved a distance of 9 m the work done is 54 J. Find the constant resistance to the motion.

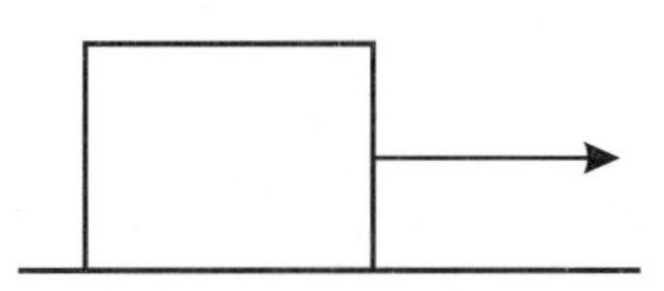

Because the speed is constant there is no acceleration.

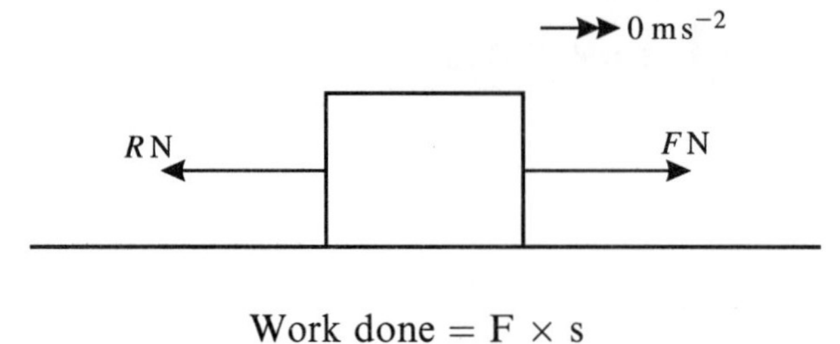

$$\text{Work done} = F \times s$$
$$54 = F \times 9$$
$$F = 6$$

Using the equation of motion: $F - R = 0$

Gives: $R = 6$

The resistance to motion is 6 N.

Work done against gravity

To raise a particle of mass m kg vertically at a constant speed you need to apply a force of mg newtons vertically upwards. If the particle is raised a distance of h metres the work done against gravity is mgh joules. Work is done against gravity when the particle is moving either vertically or at an angle to the horizontal, for example on an inclined plane, but not when the particle is moving along a horizontal plane.

Example 3

An angler catches a fish of mass 1.5 kg. Find the work he does against gravity in raising the fish 4 m from sea level to the pier, assuming he winds in his line at a constant speed.

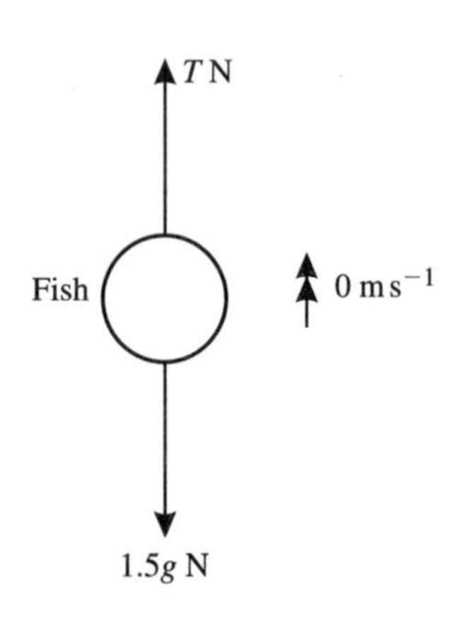

Modelling the fish as a particle moving vertically:
Since the speed is constant there is no acceleration.
Using the equation of motion vertically gives:

$$T - 1.5g = 0$$
$$T = 1.5g$$

Work done by angler against gravity $= F \times s$ and gives:

$$= 1.5 \times 9.8 \times 4$$
$$= 58.8$$

The work done by the angler is 58.8 J.

Work done against friction

When a particle is pulled up a rough inclined plane at a constant speed work must be done **against** the frictional force acting on the particle as well as against gravity (see example 4).

Example 4

A particle of mass 5 kg is pulled at constant speed a distance of 24 m up a rough plane which is inclined at 40° to the horizontal. The coefficient of friction between the particle and the surface is 0.25. Assuming the particle moves up a line of greatest slope, find:

(a) the work done against friction
(b) the work done against gravity.

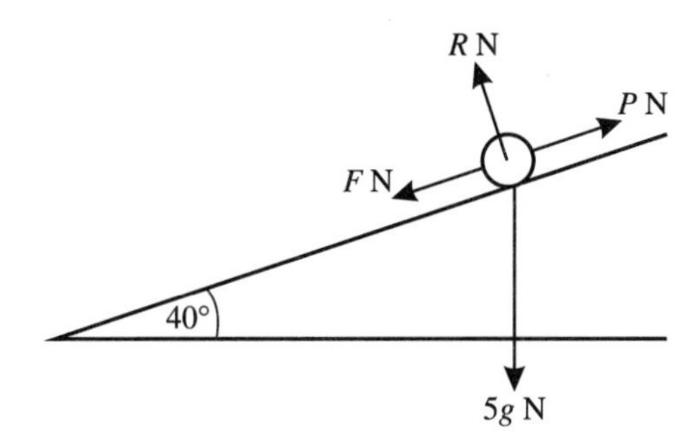

Let R be the reaction of the particle on the slope. Let F be the friction force.

(a) Because there is no motion perpendicular to the plane:

$$R - 5g\cos 40 = 0$$

Using $F = \mu R$ with $\mu = 0.25$ and $R = 5g\cos 40$ gives:

$$F = 0.25 \times 5g\cos 40$$

The particle moves a distance of 24 m parallel to the direction of the friction force.

The work done against friction is:

$$Work = F \times s$$
$$Work = F \times 24$$

Substituting the expression for F found above gives:

Work done against friction $= (0.25 \times 5g\cos 40) \times 24$
$= 225.2$

The work done against friction is 225 J.

(b) Work done against gravity = weight × vertical distance moved

$= 5g \times 24\sin 40$

$= 5 \times 9.8 \times 24 \sin 40$

$= 755.9$

The work done against gravity is 756 J.

Forces at an angle to the direction of motion

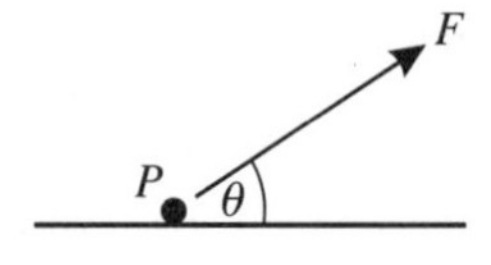

Consider a particle P resting on a horizontal surface. If a force of magnitude F inclined at an angle θ to the horizontal causes the particle P to move along the surface while remaining in contact with the surface, you can resolve the force into its horizontal and vertical components.

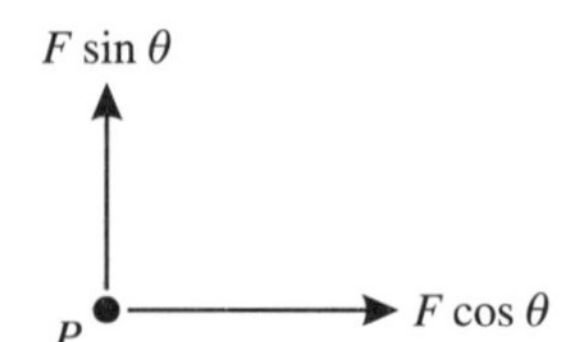

The vertical component does no work because the particle is moving in the horizontal direction only. For the horizontal component using:

$$\text{Work done} = \text{force} \times \text{distance moved}$$

Gives: $$\text{Work done} = F\cos\theta \times \text{distance moved}$$

- **For a force at an angle to the direction of motion:**
  Work done = component of force in direction of motion
  × distance moved in the same direction.

Example 5

A packing case is pulled across a smooth horizontal floor by a force of magnitude 20 N inclined at 30° to the horizontal. Find the work done by the force in moving the packing case a distance of 10 m. Model the packing case as a particle moving horizontally.

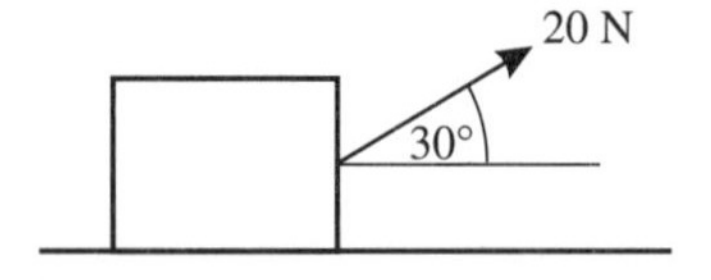

Work done = horizontal component of force × distance moved

$= 20 \cos 30 \times 10 = 173.2$

The work done is 173 J.

Exercise 3A

Whenever a numerical value of g is required take $g = 9.8\,\text{m}\,\text{s}^{-2}$.

1 Find the work done by a horizontal force of magnitude 0.7 N which pushes a particle a distance of 1.2 m across a horizontal surface.

2 A packing case is pushed 10 m across a horizontal floor by a constant horizontal force. If the work done by the force is 55 J find the magnitude of the force.

3 A package of mass 20 kg is raised 3 m vertically. By modelling the package as a particle find the work done against gravity.

4 Find the work done against gravity when a particle of mass 2 kg is raised a vertical distance of 8 m.

5 A packing case is pulled 12 m across a horizontal floor at a constant speed against resistances totalling 25 N. By modelling the packing case as a particle find the work done.

6 A cable is attached to a load of bricks of total mass 200 kg. If the work done in raising the bricks vertically at a constant speed from the ground to the top of a building is 23.5 kJ find the height of the building.

7 A packing case is pulled 20 m across a frozen pond by a rope inclined at 35° to the horizontal. The tension in the rope is 25 N. By modelling the packing case as a particle and the surface of the pond as a smooth horizontal surface, calculate the work done.

8 A particle of mass 5 kg is pulled across a rough horizontal surface. The coefficient of friction between the body and the surface is 0.4. Given that the particle moves with speed $7\,\text{m}\,\text{s}^{-1}$, find the work done against friction in 2 seconds.

9 A body of mass 5 kg is pulled a distance of 12 m across a rough horizontal surface. The body moves with constant speed and the work done against friction is 147 J. Find the coefficient of friction between the body and the surface.

10 A climber of mass 70 kg climbs a vertical cliff of height 35 m. By modelling the climber as a particle find the work he does against gravity.

11 A body of mass 2 kg is pulled at a constant speed a distance of 20 m up a smooth plane inclined at 30° to the horizontal. Find the work done against gravity.

12 A boy of mass 30 kg slides 3 m down a slide inclined at 40° to the horizontal. By modelling the boy as a particle find the work done by gravity.

13 A particle of mass 6 kg is pulled at a constant speed a distance of 15 m up a rough surface inclined at 30° to the horizontal. The coefficient of friction between the particle and the surface is 0.2. Assuming the particle moves up a line of greatest slope find the work done against friction and the work done against gravity.

14 A rough surface is inclined at arctan $\frac{3}{4}$ to the horizontal. A particle of mass 60 kg is pulled at a constant speed a distance of 40 m up a line of greatest slope of the surface. The coefficient of friction between the particle and the surface is $\frac{1}{4}$. Find (a) the magnitude of the frictional force acting on the particle (b) the work done against friction (c) the work done against gravity.

15 A rough surface is inclined at arcsin $\frac{3}{5}$ to the horizontal. A particle of mass 5 kg is pulled at a constant speed a distance of 15 m up the surface by a force acting along a line of greatest slope. The coefficient of friction between the particle and the surface is $\frac{1}{3}$. Find the total work done, assuming the only resistances to motion are due to gravity and friction.

16 A rough surface is inclined at arctan $\frac{5}{12}$ to the horizontal. A particle of mass 10 kg is pulled at a constant speed a distance of 5 m up the surface by a force acting along a line of greatest slope. The only resistances to motion are those due to gravity and friction. Given that the work done by the force is 200 J find the coefficient of friction between the particle and the surface.

3.2 Energy

The energy a particle possesses is a measure of its capability to do work. Energy exists in various forms but in this book, only kinetic energy and potential energy will be considered.

Kinetic energy

The kinetic energy of a particle is the energy it possesses by virtue of its motion.

Consider a particle of mass m on a smooth horizontal surface. A constant horizontal force F applied to the particle moves it through a distance s and increases its speed from u to v.

$$\text{Work done on the particle} = F \times s$$

The equation of motion states that $F = \textit{mass} \times \textit{acceleration}$.

Using the constant acceleration formula $v^2 = u^2 + 2as$ gives:

$$a = \frac{v^2 - u^2}{2s}$$

Substituting a into the equation of motion gives:

$$F = \frac{m(v^2 - u^2)}{2s}$$

So:

$$Fs = \tfrac{1}{2}mv^2 - \tfrac{1}{2}mu^2$$

When a particle is in motion, the expression $\frac{1}{2}$*mass* $\times$ (*velocity*)^2 is called the **kinetic energy** (or K.E.) of the particle.

$$\textbf{K.E.} = \tfrac{1}{2}mv^2$$

Since work done $= \frac{1}{2}mv^2 - \frac{1}{2}mu^2$:

work done by a force = K.E. at end – K.E. at start

or: work done = increase in K.E.

If more than one force acts on the particle, for example a particle being pulled along a rough surface, then the force used to calculate the work done is the resultant force in the direction of the motion.

Because work done = increase in K.E., work and kinetic energy are measured in the same units.

For a mass in kilograms and a velocity in metres per second the kinetic energy is measured in joules.

Example 6

A particle of mass 0.5 kg is moving with a speed of $6\,\text{m s}^{-1}$. Find its kinetic energy.

You know that: $\text{K.E.} = \frac{1}{2}mv^2$

Substituting known quantities gives: $\text{K.E.} = \frac{1}{2} \times 0.5 \times 6^2$

$= 9$

The kinetic energy of the particle is 9 J.

Example 7

A particle of mass 2 kg is being pulled across a smooth horizontal surface by a horizontal force. The force does 24 J of work in increasing the particle's velocity from $5\,\text{m s}^{-1}$ to $v\,\text{m s}^{-1}$. Find the value of v.

$$\text{Work done} = \tfrac{1}{2}mv^2 - \tfrac{1}{2}mu^2$$

Initial K.E. of particle is:

$$\tfrac{1}{2}mu^2 = \tfrac{1}{2} \times 2 \times 5^2$$

So:

$$24 = \tfrac{1}{2}mv^2 - \tfrac{1}{2} \times 2 \times 5^2$$

$$\tfrac{1}{2}mv^2 = 24 + 25$$

$$\tfrac{1}{2} \times 2v^2 = 49$$

$$v^2 = 49$$

$$v = 7$$

The final speed of the particle is $7\,\text{m s}^{-1}$.

Example 8

A car of mass 1200 kg starts from rest at a set of traffic lights. After travelling 300 m its speed is $20\,\text{m}\,\text{s}^{-1}$. Given that the car is subject to a constant resistance of 400 N find the driving force, assumed constant.

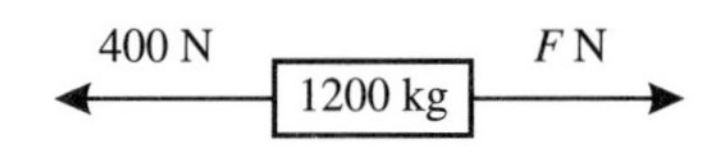

$$\text{Increase in K.E.} = \tfrac{1}{2}mv^2 - \tfrac{1}{2}mu^2$$
$$= \tfrac{1}{2} \times 1200 \times 20^2 - 0$$

Resultant force in direction of motion $= F - 400$.

Because: Work done = increase in K.E.

and: Work done = force × distance moved

Then:
$$(F - 400) \times 300 = \tfrac{1}{2} \times 1200 \times 20^2$$
$$F - 400 = \frac{\tfrac{1}{2} \times 1200 \times 20^2}{300}$$
$$F = 800 + 400 = 1200$$

The driving force is 1200 N.

Potential energy

The **potential energy** of a particle is the energy it possesses by virtue of its position. As shown earlier on page 60, when a particle of mass m kg is *raised* through a vertical distance h metres the work done against gravity is mgh joules. This work is equal to the *increase* in potential energy (or P.E.) of the particle. If the particle is *lowered* through a vertical distance h' then its potential energy is *decreased* by mgh' joules. For each situation choose a zero level for P.E. and indicate this on the diagram. Changes in potential energy can then be calculated relative to that level.

■ **P.E. =** $\boldsymbol{mgh}$

Example 9

A parcel of mass 5 kg is raised vertically through a distance of 2 m. Find the increase in potential energy.

Model the parcel as a particle moving vertically upwards and take the original level of the parcel as the zero level for P.E.

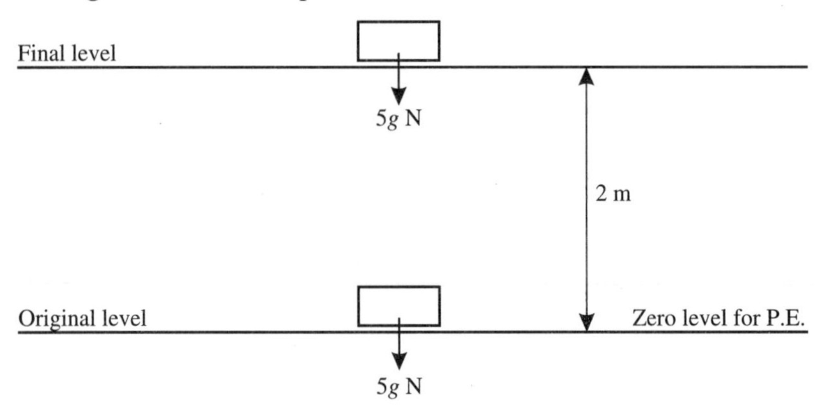

Initial P.E. $= 0$

Final P.E. $= mgh = 5 \times 9.8 \times 2 = 98$

The increase in P.E. is 98 J.

Example 10

A child of mass 25 kg slides 4 m down a playground slide inclined at an angle of $\arcsin \frac{3}{5}$ to the horizontal. Model the child as a particle and the slide as an inclined plane and hence calculate the potential energy lost by the child.

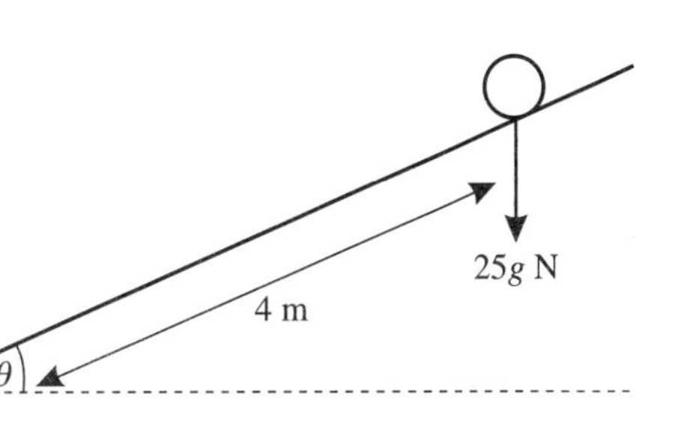

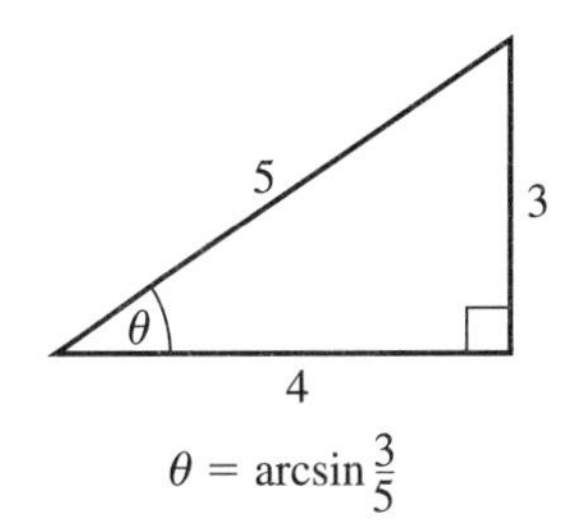

Take the bottom of the slide to be the zero level for P.E.

Vertical distance moved by the child $= 4 \sin \theta = 4 \times \frac{3}{5}$

Initial P.E. $= mgh = 25 \times 9.8 \times 4 \times \frac{3}{5} = 588$

Final P.E. $= 0$

The potential energy lost by the child is 588 J.

Exercise 3B

Whenever a numerical value of g is required take $g = 9.8 \text{ m s}^{-2}$.

1 Find the kinetic energy of:

(a) a particle of mass 0.2 kg moving at 10 m s^{-1}

(b) a particle of mass 2 kg moving at 2 m s^{-1}

(c) a package of mass 10 kg moving at 5 m s^{-1}

(d) a bullet of mass 15 g moving at 500 m s^{-1}

(e) a car of mass 750 kg moving at 17 m s^{-1}

(f) a man of mass 65 kg running at 5 m s^{-1}.

2 Find the change in potential energy of each of the following, stating in each case whether it is a loss or a gain:

(a) a particle of mass 0.5 kg raised through a vertical distance of 7 m

(b) a man of mass 65 kg descending a vertical distance of 20 m

(c) a lift of mass 600 kg ascending a vertical distance of 40 m

(d) a man of mass 70 kg ascending a vertical distance of 100 m

(e) a bucket of water of mass 10 kg raised a vertical distance of 22 m

(f) a lift of mass 900 kg descending a vertical distance of 30 m.

3 A particle of mass 0.5 kg increases its speed from $5\,\mathrm{m\,s^{-1}}$ to $7\,\mathrm{m\,s^{-1}}$. Find the increase in the kinetic energy of the particle.

4 A car of mass 750 kg decreases speed from $30\,\mathrm{m\,s^{-1}}$ to $5\,\mathrm{m\,s^{-1}}$. Find the decrease in the kinetic energy of the car.

5 A particle of mass 0.5 kg increases its kinetic energy by 15 J. Its initial speed was $2\,\mathrm{m\,s^{-1}}$. Find its final speed.

6 A boy of mass 40 kg initially running at $4\,\mathrm{m\,s^{-1}}$ decreases his kinetic energy by 140 J. Find his final speed.

7 A child of mass 30 kg slides 5 m down a playground slide inclined at 40° to the horizontal. By modelling the child as a particle calculate the potential energy lost by the child.

8 A particle of mass 5 kg is moving at $2\,\mathrm{m\,s^{-1}}$ when it is acted on by a force of magnitude 10 N for 2 seconds, causing it to accelerate. Find the increase in kinetic energy.

9 A car of mass 750 kg initially travelling at $40\,\mathrm{m\,s^{-1}}$ retards at $2\,\mathrm{m\,s^{-2}}$ for 4 s. By modelling the car as a particle calculate the kinetic energy lost.

10 A stone of mass 0.8 kg is dropped from a height of 1.5 m into a pond. Find the kinetic energy of the stone as it hits the surface of the water. The stone begins to sink in the water with a speed of $2.1\,\mathrm{m\,s^{-1}}$. Determine the kinetic energy lost when the stone strikes the water.

Conservation of energy

The only force acting on a particle of mass m falling freely under gravity is its weight. If the particle increases its speed from $u\,\mathrm{m\,s^{-1}}$ to $v\,\mathrm{m\,s^{-1}}$ while descending a distance h metres then:

$$\text{Work done by gravity} = mgh$$

And: $\quad$ Increase in K.E. of particle = final K.E. − initial K.E.

$$= \tfrac{1}{2}mv^2 - \tfrac{1}{2}mu^2$$

Since: $\quad$ Work done = increase in K.E.

It follows that: $\quad mgh = \tfrac{1}{2}mv^2 - \tfrac{1}{2}mu^2$

But as seen earlier the work done by gravity, mgh, is also equal to the decrease in potential energy.

$$mgh = \text{decrease in P.E.} = \text{initial P.E.} - \text{final P.E.}$$

So: $\quad$ decrease in P.E. = increase in K.E.

or: $\quad$ initial P.E. − final P.E. = final K.E. − initial K.E.

And so: $\quad$ initial (P.E. + K.E.) = final (K.E. + P.E.)

- **A particle's total energy is constant, if it is subject only to gravity.**

This is true whether the particle moves vertically or along a path inclined to the vertical and is an application of the **Principle of conservation of energy**.

Example 11

A particle of mass 2 kg is released from rest and slides down a smooth plane inclined at $\arcsin \frac{3}{5}$ to the horizontal. Find the distance travelled while the particle increases its velocity to $5\,\text{m}\,\text{s}^{-1}$.
Let the distance travelled be x m.

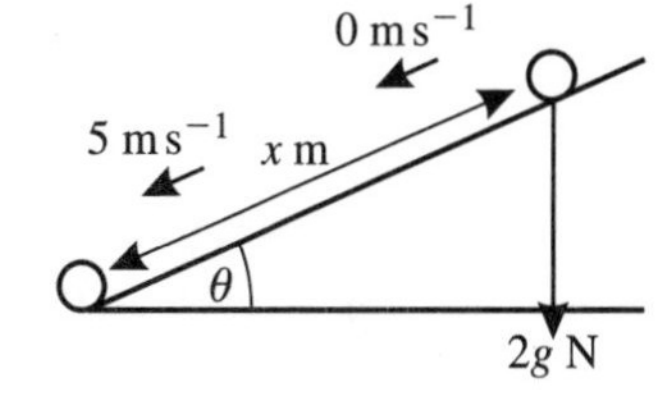

Then:

$$\begin{aligned}\text{Decrease in height} &= x \sin\theta \\ &= \tfrac{3}{5}x \\ \text{Decrease in P.E.} &= mgh \\ &= 2 \times 9.8 \times \tfrac{3}{5}x \\ \text{Increase in K.E.} &= \tfrac{1}{2}mv^2 - \tfrac{1}{2}mu^2 \\ &= \tfrac{1}{2} \times 2 \times 5^2 - 0\end{aligned}$$

Since decrease in P.E. = increase in K.E:

$$2 \times 9.8 \times \tfrac{3}{5}x = \tfrac{1}{2} \times 2 \times 5^2$$

$$x = \frac{5 \times 5^2}{2 \times 9.8 \times 3} = 2.125$$

The distance travelled is 2.13 m.

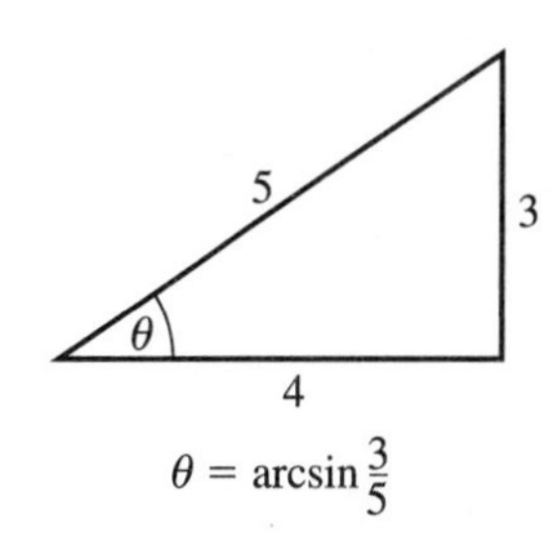

$\theta = \arcsin \frac{3}{5}$

Change of energy of a particle

If the energy possessed by a particle changes during the motion being considered it must be as a result of some force doing work on that particle.

If a resistance is acting on the particle, the resistance is working against the motion and will cause a loss of energy.

Example 12

A particle of mass 5 kg is projected up a rough plane inclined at arcsin $\frac{4}{5}$ to the horizontal with a speed of 6 m s^{-1}. Given that the coefficient of friction between the particle and the plane is $\frac{1}{3}$ find the distance the particle moves up the plane before coming to rest.

Let the distance travelled up the plane be x metres.

Let F N be the friction force and R N the normal reaction.

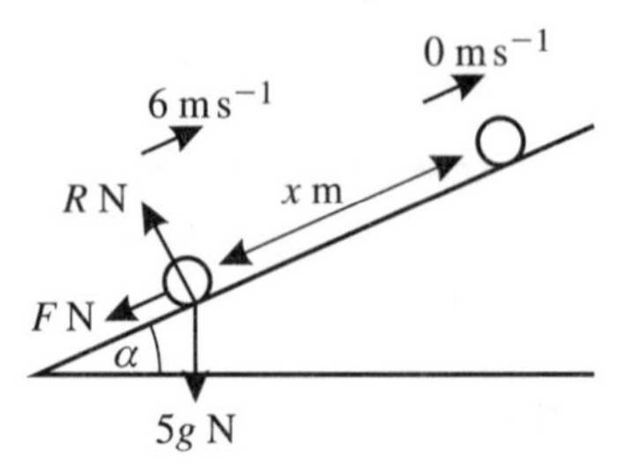

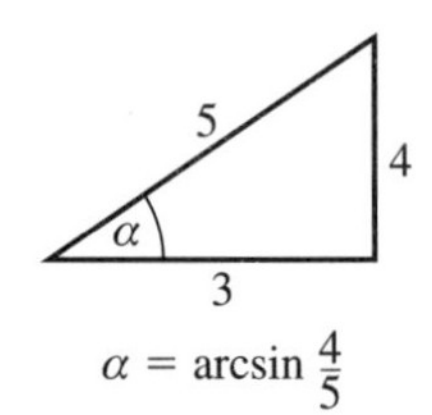

$\alpha = \arcsin \frac{4}{5}$

The particle comes to rest because of the work done by friction.

Initial K.E. $= \frac{1}{2}mv^2 = \frac{1}{2} \times 5 \times 6^2 = 90$

Final K.E. $= 0$

P.E. gained $= mgh$

$$= 5 \times 9.8 \times x \sin \alpha$$
$$= 5 \times 9.8 \times \tfrac{4}{5}x$$
$$= 39.2x$$

Total loss of energy = K.E. lost − P.E. gained

$$= (90 - 39.2x)$$

So total loss of energy is $(90 - 39.2x)$ J

Work done by friction $= F \times x$

Since there is no motion perpendicular to the plane:

$$R - 5g \cos \alpha = 0$$
$$R = 5g \cos \alpha = 5g \times \tfrac{3}{5} = 3g$$

Because the particle is moving up the plane $F = \mu R$.

Therefore: $\quad F = \frac{1}{3} \times 3g = g = 9.8$

So work done by friction is $(9.8x)$ J.

Since: $\quad$ Work done by friction = loss of energy

$$9.8x = 90 - 39.2x$$
$$x = \frac{90}{49} = 1.84$$

The distance moved is 1.84 m.

The above example can also be solved by using the equation of motion and the uniform acceleration equations. However, an examination question may specify 'by considering the energy of the particle' or similar wording in which case the above method must be used.

Example 13

A cyclist reaches the top of a hill with a speed of $4\,\text{m s}^{-1}$. He descends 40 m and then ascends 35 m to the top of the next incline. His speed is now $3\,\text{m s}^{-1}$. The cyclist and his bicycle have a combined mass of 90 kg. The total distance he cycles from the top of the first hill to the top of the next incline is 750 m and there is a constant resistance to motion of 15 N. Find the work done by the cyclist.

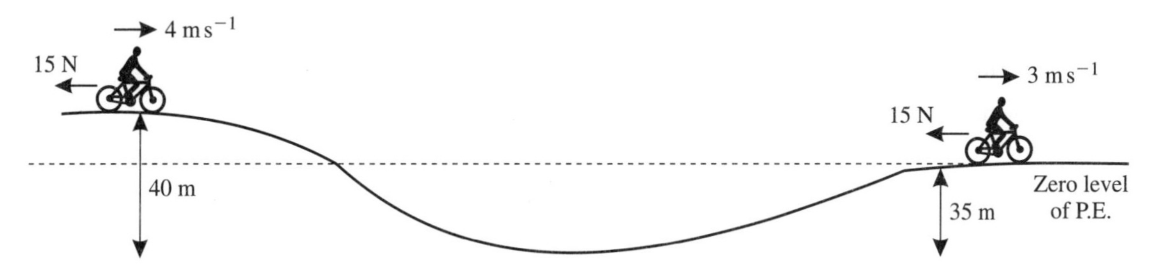

Take the cyclist's final level as the zero level for P.E.

Loss of K.E. $= \frac{1}{2}mu^2 - \frac{1}{2}mv^2$

$= \left(\frac{1}{2} \times 90 \times 4^2 - \frac{1}{2} \times 90 \times 3^2\right)$

$= 315$

Loss of P.E. $= mgh$

$= 90 \times 9.8 \times (40 - 35)$

$= 4410$

Total loss of energy is 4725 J.

Work done against resistances $F \times s = (15 \times 750) = 11\,250$ J.

The cyclist provides the work done against the resistances but he is 'helped' by his own loss of energy.

Work done by the cyclist = work done against the resistances − energy lost

$= (11\,250 - 4725) = 6525$

Thus work done by the cyclist is 6525 J.

Exercise 3C

Whenever a numerical value of g is required take $g = 9.8\ \text{m s}^{-2}$.

1 A particle of mass 0.6 kg falls a vertical distance of 5 m from rest. Neglecting resistances, find the potential energy lost and hence the final speed of the particle.

2 A stone of mass 1 kg is dropped from the top of a vertical cliff. If it hits the ground with a speed of $20\,\text{m s}^{-1}$ find (a) the kinetic energy gained (b) the potential energy lost (c) the height of the cliff.

3 A package of mass 8 kg is pushed in a straight line across a smooth horizontal floor by a constant horizontal force of magnitude 16 N. The package has speed $3\,\text{m s}^{-1}$ when it passes through point A and speed $5\,\text{m s}^{-1}$ when it reaches point B. Find (a) the increase in kinetic energy of the package (b) the work done by the force (c) the distance AB.

4 A package of mass 6 kg is moving in a straight line across a rough horizontal surface. The coefficient of friction between the package and the surface is $\frac{1}{3}$. The package passes through point A with a speed of $5\,\text{m s}^{-1}$ and comes to rest at B. Find (a) the kinetic energy lost by the package (b) the work done against friction (c) the distance AB.

5 A particle of mass 0.2 kg slides in a straight line across a rough horizontal surface. The speed of the particle decreases from $10\,\text{m s}^{-1}$ to $5\,\text{m s}^{-1}$ while travelling 9 m. Find (a) the kinetic energy lost (b) the work done against friction (c) the coefficient of friction between the particle and the surface.

6 A ball of mass 0.5 kg falls a vertical distance of 7 m from rest. By considering energy, find the speed of the ball when it hits the ground.

7 A stone of mass 0.2 kg is dropped by a bird which is flying horizontally. If the stone hits the horizontal ground with a speed of $40\,\text{m s}^{-1}$ by considering energy, find how high above the ground the bird is flying.

8 A bullet of mass 10 g travelling at $400\,\text{m s}^{-1}$ horizontally hits a vertical wall and penetrates the wall to a depth of 40 mm. Assuming the resistive force exerted by the wall to be constant find the magnitude of the force.

9 A bullet of mass 20 g travelling at $500\,\text{m s}^{-1}$ horizontally hits a vertical wall. The wall exerts a constant resistance of 36 000 N on the bullet. Calculate the distance the bullet will penetrate the wall.

10 A body of mass 3 kg is released from rest and slides 2 m down a smooth plane inclined at 40° to the horizontal. Find (a) the potential energy lost (b) the kinetic energy gained (c) the final speed of the body.

11 A body of mass 5 kg is released from rest and slides down a smooth plane inclined at 30° to the horizontal. Find the distance travelled while increasing its speed to $4\,\text{m s}^{-1}$.

12 A particle of mass 0.4 kg is projected up a line of greatest slope of a smooth plane inclined at 35° to the horizontal with a velocity of $12\,\text{m}\,\text{s}^{-1}$. Use energy considerations to find the distance the particle travels before first coming to rest.

13 A particle of mass 0.5 kg is projected up a line of greatest slope of a smooth plane inclined at arcsin $\frac{3}{5}$ to the horizontal. Given that the particle travels a distance of 10 m before first coming to rest find the speed of projection.

14 A particle of mass 4 kg is projected up a rough plane inclined at arcsin $\frac{5}{13}$ to the horizontal with a velocity of $8\,\text{m}\,\text{s}^{-1}$. Given that the coefficient of friction between the single particle and the plane is $\frac{1}{4}$ find, by energy considerations, the distance the particle moves up the plane before coming to rest.

15 A cyclist and his machine have a combined mass of 90 kg. The cyclist freewheels down a hill inclined at 30° to the horizontal. He increases his speed from $5\,\text{m}\,\text{s}^{-1}$ to $25\,\text{m}\,\text{s}^{-1}$. By modelling the cyclist and his bicycle as a single particle and assuming resistances can be neglected, calculate the potential energy lost by the cyclist and the distance he travels.

16 A cyclist starts from rest and freewheels down a hill inclined at arcsin $\frac{1}{20}$ to the horizontal. After travelling 80 m the road becomes horizontal and the cyclist travels another 80 m before coming to rest without using his brakes. Given that the combined mass of the cyclist and his machine is 80 kg, model the cyclist and his machine as a single particle and hence find the resistive force, assumed constant throughout the motion.

17 A boy and his skateboard have a mass of 50 kg. He descends a slope inclined at 12° to the horizontal starting from rest. At the bottom, the ground becomes horizontal for 10 m before rising at 8° to the horizontal. The boy travels 30 m up the incline before coming to rest again. He is subject to a constant resistance of 20 N throughout the motion. By modelling the boy and his skateboard as a single particle find the distance the boy travelled down the slope.

18 A tile slides down a smooth roof inclined at 45° to the horizontal and off the edge which is 10 m above the ground. Given that the tile started from rest 5 m from the edge of the roof find the speed with which it hits the ground.

19 A book slides down a rough desk lid which is inclined at 30° to the horizontal and falls to the ground which is 0.8 m below the edge of the lid. The coefficient of friction between the book and the lid is $\frac{1}{4}$. Given that the book starts from rest 0.4 m from the bottom of the lid which is hinged at the edge of the desk, find the speed with which the book hits the floor.

20 A rough plane is inclined at $\arcsin \frac{3}{5}$ to the horizontal. A package of mass 40 kg is released from rest and travels 15 m while increasing speed to $12\,\text{m}\,\text{s}^{-1}$. Find, using energy considerations, the coefficient of friction between the package and the plane.

3.3 Power

Power is defined as the rate of doing work. It is measured in watts (W) where **1 watt is 1 joule per second**. The power of an engine is frequently given in kilowatts (kW). **1 kilowatt is 1000 watts**. An engine which develops a power of 1 kilowatt is doing 1000 joules of work every second.

Example 14

A force of magnitude 1500 N pulls a truck up a slope at a constant speed of $6\,\text{m}\,\text{s}^{-1}$. Given that the force acts parallel to the direction of motion find in kW the power developed.

Work done per second = force × distance moved in 1 second

$$= (1500 \times 6) = 9000\,\text{J}$$

Power = rate of doing work = 9000 W

The power developed is 9 kW.

Moving vehicles

If the engine of a vehicle is producing a driving force of F newtons when the vehicle has a speed of $v\,\text{m}\,\text{s}^{-1}$ then the work done per second is $F \times$ distance moved in one second $= F \times v$.

Because work done per second is the power developed by the engine:

■ **Power = $F \times v$**

Example 15

A car of mass 1000 kg is travelling along a level road against a constant resistance of magnitude 475 N. The engine of the car is working at 4 kW. Calculate:

(a) the acceleration when the car is travelling at $5\,\text{m}\,\text{s}^{-1}$
(b) the maximum speed of the car.

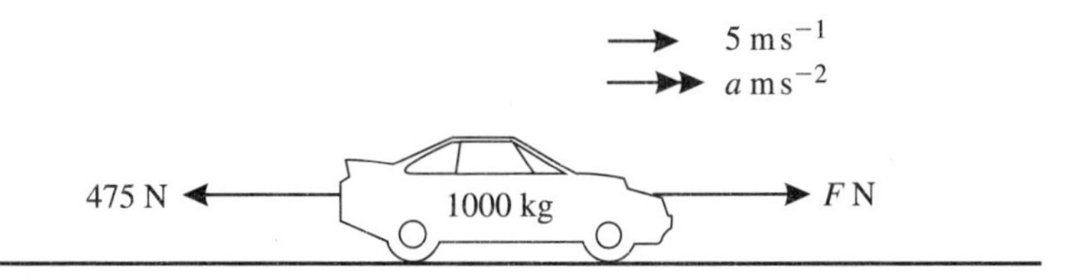

(a) Power = 4 kW = 4000 W

Since: $\quad \text{Power} = F \times v$

So: $\quad 4000 = F \times 5$

$$F = 800$$

Using $F = ma$ where F is the resultant force gives:

$$800 - 475 = 1000a$$

$$a = \frac{800 - 475}{1000} = 0.325$$

The acceleration is $0.325\,\text{m}\,\text{s}^{-2}$.

(b) When the car is travelling at its maximum speed there will be no acceleration so the resultant force in the direction of motion will be zero.

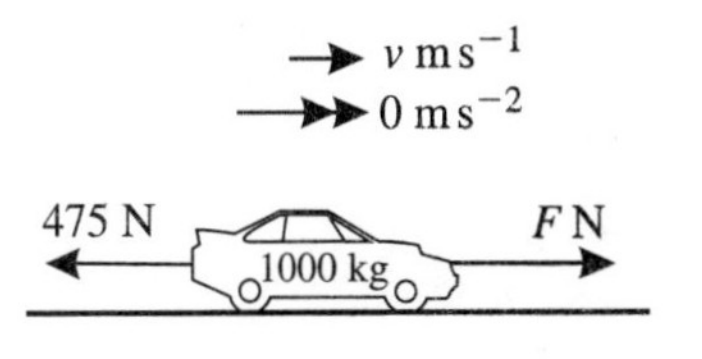

Hence: $\quad F = 475N$

Using: $\quad \text{Power} = F \times v$

Gives: $\quad 4000 = 475v$

$$v = \frac{4000}{475} = 8.421$$

The maximum speed of the car is $8.42\,\text{m}\,\text{s}^{-1}$.

Example 16

A car of mass 1200 kg is moving up a hill of slope $\arcsin \frac{1}{15}$ at a constant speed of $20\,\text{m}\,\text{s}^{-1}$. If the power developed by the engine is 25 kW find the resistance to motion.

At the top of the hill the road becomes horizontal. Find the initial acceleration, assuming the resistance to be unchanged.

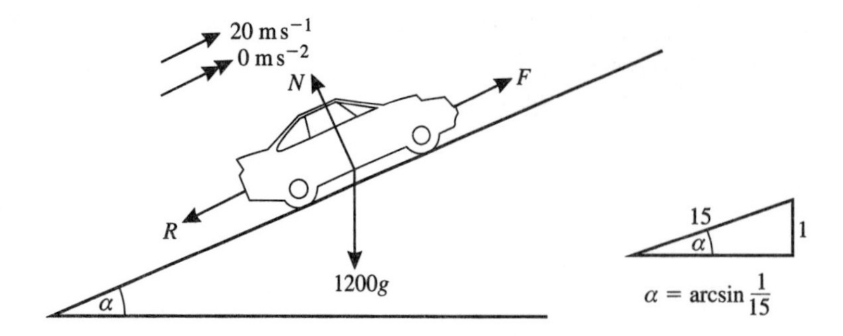

The driving force produced by an engine is often called the **tractive force.**

Let the tractive force be F.

Because:

$$\text{Power} = F \times v$$

$$25 \times 10^3 = F \times 20$$

$$F = \frac{25 \times 10^3}{20}$$

$$= 1250$$

Since the speed is constant, there is no acceleration, and so there is no resultant force in the direction of motion. Hence:

$$R + 1200g \sin \theta = F$$

$$= 1250$$

So:

$$R = 1250 - 1200 \times 9.8 \times \frac{1}{15}$$

$$= 466$$

The resistance to motion is 466 N.

At the instant the road becomes horizontal, the tractive force is 1250 N.

Using:

$$F = ma$$

Gives:

$$1250 - 466 = 1200a$$

$$a = \frac{1250 - 466}{1200} = 0.6533$$

a m s^{-2}

466 N 1250 N

The initial acceleration when the car travels along the horizontal part of the road is 0.653 m s^{-2}.

In the above example, as the car's speed increases either the power which equals force × velocity, must increase or the driving force must decrease and there will be a corresponding decrease in acceleration.

Exercise 3D

Whenever a numerical value of g is required take $g = 9.8\,\text{m}\,\text{s}^{-2}$.

1 A force of 1200 N pulls a truck up a slope at a constant speed of $10\,\text{m}\,\text{s}^{-1}$. Find in kW the power developed.

2 The engine of a car is producing a driving force of 800 N. Given that the car is travelling at $12\,\text{m}\,\text{s}^{-1}$ find the power developed.

3 The engine of a car is working at 4 kW. Given that the car is travelling at $15\,\text{m}\,\text{s}^{-1}$ find the driving force produced by the engine.

4 A car is travelling along a level road against a constant resistance of 500 N. The engine of the car is working at 4.5 kW. Calculate the maximum speed of the car.

5 A car is travelling along a level road against a constant resistance of 475 N. The engine of the car is working at 6 kW. Calculate the maximum speed of the car.

6 A car is travelling along a level road at a constant speed of $15\,\text{m}\,\text{s}^{-1}$. The engine of the car is working at 7.5 kW. Calculate the resistance to motion.

7 A car of mass 900 kg is travelling along a level road against a constant resistance of 400 N. The engine of the car is working at 8 kW. Calculate (a) the acceleration when the car is travelling at $5\,\text{m}\,\text{s}^{-1}$ (b) the acceleration when the car is travelling at $12\,\text{m}\,\text{s}^{-1}$ (c) the maximum speed of the car.

8 A car of mass 1000 kg is travelling along a level road against a constant resistance. When the engine is working at 12 kW and the speed of the car is $15\,\text{m}\,\text{s}^{-1}$, the acceleration is $0.4\,\text{m}\,\text{s}^{-2}$. Find the resistive force.

9 A car of mass 800 kg is travelling along a level road at a speed of $20\,\text{m}\,\text{s}^{-1}$. The car is subject to a constant resistance of 250 N and its acceleration is $0.25\,\text{m}\,\text{s}^{-2}$. Find the power developed by the engine.

10 A cyclist is travelling along a level road against a constant resistance of 30 N. The maximum rate at which the cyclist can work is 300 W. Find her maximum speed.

11 A car of mass 1000 kg is moving up a hill inclined at arcsin $\frac{1}{20}$ at a steady speed of $25\,\text{m}\,\text{s}^{-1}$. Given that the power developed by the engine is 20 kW find the resistance to motion. The road now becomes horizontal. Find the initial acceleration, assuming the resistance to be unchanged.

12 A car of mass 1000 kg is travelling along a level road against a constant resistance of 400 N. Given that the maximum speed of the car is $35\,\text{m}\,\text{s}^{-1}$ find the power developed by the engine. With the engine of the car working at the same rate and the resistance unchanged the car ascends a slope inclined at $10°$ to the horizontal. Find its maximum velocity up the hill.

13 The mass of a car is 750 kg. Its engine is working at a constant rate of 15 kW against a constant resistance of magnitude 600 N. Given that the car is travelling at $20\,\text{m}\,\text{s}^{-1}$ along a horizontal road find the acceleration of the car. The car now ascends a straight road, inclined at $6°$ to the horizontal. Given that the car's engine works at the same rate and the resistance to motion is unchanged find the maximum speed of the car up the slope.

14 A train of mass 200 tonnes is ascending a hill inclined at $0.5°$ to the horizontal. The engine is working at a constant rate of 300 kW and the resistances to motion amount to 8 kN. Find the maximum speed of the train. The track now becomes horizontal. Given that the engine works at the same rate and the resistances to motion are unchanged find the initial acceleration of the train.

15 The resistance to motion of a car moving with speed $v\,\text{m}\,\text{s}^{-1}$ is given by the formula $(200 + 2v)$ N. Given that the engine of the car is working at 8 kW find the maximum speed of the car as it travels along a horizontal road.

Exercise 3E Mixed questions

Whenever a numerical value of g is required take $g = 9.8\,\text{m}\,\text{s}^{-2}$.

1 A car of mass 750 kg is travelling at a speed of $12\,\text{m}\,\text{s}^{-1}$ along a straight road against a constant resistance of magnitude 420 N. The engine of the car is working at a rate of 15 kW. In an initial model of the situation the road is assumed to be horizontal.

(a) Calculate the acceleration of the car.

In a refined model the road is assumed to be inclined at an angle of 5° to the horizontal. No other changes are made.

(b) Calculate the revised acceleration.

2 A box of mass 10 kg is resting on a sloping platform which is inclined at an angle of 20° to the horizontal. The coefficient of friction between the box and the platform is 0.2. The box is released from rest and slides down the platform. Calculate:

(a) the speed of the box after it has been moving for 3 s

(b) the potential energy lost by the box during this time.

3 A lorry of mass 15 000 kg is moving up a straight road inclined at an angle θ to the horizontal where $\sin\theta = 0.25$. The lorry is travelling at a constant speed of $12\,\text{m s}^{-1}$ and the constant resistance to its motion has magnitude 175 kN. Find the work done in 10 s by the engine of the lorry.

4 A particle P of mass 0.5 kg is moving in a straight line on a smooth horizontal surface under the action of a constant horizontal force. The particle passes point A with speed $4\,\text{m s}^{-1}$ and passes point B with speed $10\,\text{m s}^{-1}$. Find:

(a) the kinetic energy gained by P while moving from A to B

(b) the work done by the constant force.

The distance from A to B is 25 m.

(c) Calculate the magnitude of the force.

5 A car of mass 1400 kg is towing a caravan of mass 400 kg up a straight road which is inclined at an angle θ to the horizontal, where $\sin\theta = 0.15$. The engine of the car is working at a constant rate of 60 kW. The resistances acting on the car and caravan have magnitude 800 N and 300 N respectively. The tow-bar is light and inextensible and can be assumed to be parallel to the road. At the instant when the speed of the car is $12\,\text{m s}^{-1}$ calculate:

(a) the magnitude of the acceleration of the car

(b) the tension in the tow-bar.

6 A car of mass 1200 kg is moving up a straight road which is inclined to the horizontal at 15°. The car experiences a constant resistance of magnitude 300 N. Calculate the power of the engine in kW at the instant when the car has an acceleration $0.75\,\text{m s}^{-2}$ and is moving at $12\,\text{m s}^{-1}$.

7

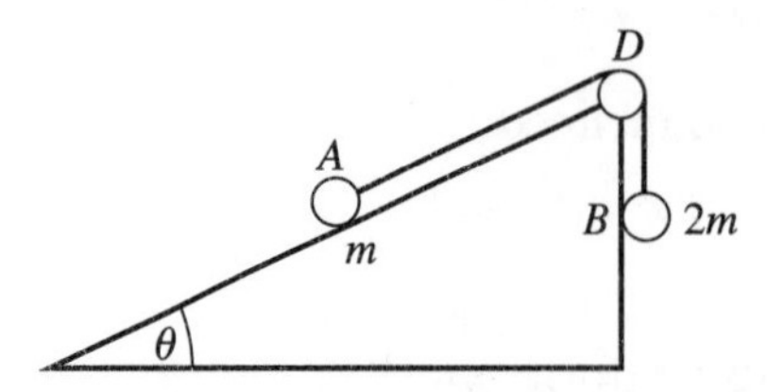

The diagram shows a particle A of mass m which can move on the rough surface of a plane inclined at an angle θ to horizontal ground, where $\theta = \arcsin 0.6$. A second particle B of mass $2m$ hangs freely attached to a light inextensible string which passes over a smooth pulley fixed at D. The other end of the string is attached to A. The coefficient of friction between A and the plane is $\frac{1}{4}$. B is initially hanging 3 m above the ground and A is 2 m from D. When the system is released from rest with the string taut A moves up the plane.

(a) Find the initial acceleration of A.

When B has descended 1 m the string breaks. By using the principle of conservation of energy

(b) calculate the total distance moved by A before it first comes to rest.

8 A particle P of mass 0.2 kg is at rest 3 m from the edge of a horizontal bench. P receives an impulse of magnitude 4 N s.

(a) Calculate the speed with which P begins to move.

The bench is rough and the coefficient of friction between P and the bench is $\frac{2}{5}$. Calculate:

(b) the work done against friction as P moves from its initial position to the edge of the bench

(c) the speed of P when P reaches the edge of the bench.

The surface of the bench is 1 m above a horizontal floor.

(d) Calculate the speed with which P hits the floor.

9 A parcel of mass 5 kg rests on a rough horizontal floor. The coefficient of friction between the parcel and the floor is 0.2. The parcel is pulled across the floor by a rope which is inclined at 10° to the horizontal. The tension in the rope is constant and has magnitude 50 N.

(a) Find the frictional force opposing the motion.

(b) Find the work done by the tension when moving the parcel a distance of 10 m.

The parcel is initially at rest.

(c) Find the speed of the parcel when it has moved 10 m.

10 The engine of a car works at a constant rate of 16 kW. The car has a mass of 1500 kg and on a straight level road there is a constant resistance to motion of 400 N.

(a) Determine the maximum speed of the car, in m s^{-1}.

(b) Find the acceleration of the car, in m s^{-2}, when its speed is $25\,\text{m s}^{-1}$.

SUMMARY OF KEY POINTS

1 **Work**

For a force acting in the direction of the motion:

$$\text{Work done} = \text{force} \times \text{distance moved}$$

For a force acting in a direction other than that of the motion:

Work done = component of force in the direction of motion × distance moved in the same direction

A force of 1 newton (N) does 1 joule (J) of work when moving a particle a distance of 1 metre.

2 **Energy**

The kinetic energy (K.E.) of a body of mass m moving with speed $v\,\text{m s}^{-1}$ is given by:

$$\text{K.E.} = \tfrac{1}{2}mv^2$$

For a mass in kg and velocity in m s^{-1} the K.E. is measured in joules (J). K.E. is never negative.

The potential energy (P.E.) of a body of mass m at a height h above a chosen level is given by:

$$\text{P.E.} = mgh$$

Potential energy is also measured in joules. P.E. can be negative.

The work done on a body is equal to its change of mechanical energy which is K.E. + P.E.

3 **Conservation of energy**

If the weight of the particle is the only force having a component in the direction of motion then throughout the motion:

$$\text{K.E.} + \text{P.E.} = \text{constant}$$

4 **Power**

Power is the rate of doing work.

For a moving particle:

$$\text{Power} = \text{driving force} \times \text{speed}$$

Power is measured in watts, where 1 watt (W) is 1 joule per second, or kilowatts, where 1 kilowatt (kW) is 1000 watts.

Review exercise 1

Whenever a numerical value of g is required take $g = 9.8 \text{ m s}^{-2}$.

1 A stone is thrown from a point 12.6 m above horizontal ground with speed 7 m s^{-1} at an angle of $30°$ above the horizontal. Calculate the time which elapses before the stone hits the ground. [E]

2 A bowler delivers a ball horizontally at 30 m s^{-1}, his hand being 2 m above the horizontal ground at the instant that the ball is released. Find, in metres to the nearest metre, the horizontal distance that the ball travels before striking the ground. [E]

3 At time $t = 0$ a ball is kicked from a point A on horizontal ground and moves freely under gravity. At time $t = 4$ s the ball hits the ground at B, where $AB = 60$ m. Calculate the horizontal and vertical components of the velocity of the ball as it leaves A. [E]

4 A ball was thrown from a balcony above a horizontal lawn. The velocity of projection was 10 m s^{-1} at an angle of elevation α, where $\tan \alpha = \frac{3}{4}$. The ball moved freely under gravity and took 3 s to reach the lawn from the instant when it was thrown. Calculate:

(a) the vertical height above the lawn from which the ball was thrown

(b) the horizontal distance between the point of projection and the point A at which the ball hit the lawn

(c) the angle, to the nearest degree, between the direction of the velocity of the ball and the horizontal at the instant when the ball reached A. [E]

5 The unit vectors **i** and **j** are horizontal and vertically upwards respectively. A particle is projected with velocity $(8\mathbf{i} + 10\mathbf{j}) \text{ m s}^{-1}$ from a point O at the top of a cliff and moves freely under gravity. Six seconds after projection, the particle strikes the sea at the point S. Calculate:

(a) the horizontal distance between O and S

(b) the vertical distance between O and S.

(c) At time T seconds after projection, the particle is moving with velocity $(8\mathbf{i} - 14.5\mathbf{j})\,\text{m s}^{-1}$. Find the value of T and the position vector, relative to O, of the particle at this instant. [E]

6 A golf ball is hit from a point O on horizontal ground and moves freely under gravity. The horizontal and vertical components of the initial velocity of the ball are $2u\,\text{m s}^{-1}$ and $u\,\text{m s}^{-1}$ respectively. The ball hits the horizontal ground at a point whose distance from O is 80 m.

(a) Show that $u = 14$.

(b) Show that the highest point of the path of the ball is 10 m above the ground.

(c) Find, in m s^{-1} to 2 significant figures, the speed of the ball $2\frac{1}{2}$ seconds after the ball has been hit.

(d) State four assumptions you have made about the ball and the forces acting on it during its flight. [E]

7 A pilot, flying an aeroplane in a straight line at a constant speed of $196\,\text{m s}^{-1}$ and at a constant height of 2000 m, drops a bomb on a stationary ship in the vertical plane through the line of flight of the aeroplane. Assuming that the bomb falls freely under gravity, calculate:

(a) the time which elapses after release before the bomb hits the ship

(b) the horizontal distance between the aeroplane and the ship at the time of release of the bomb

(c) the speed of the bomb just before it hits the ship. [E]

8 A particle P is projected, from a point O on horizontal ground, with speed $V\,\text{m s}^{-1}$ and angle of elevation α. The particle moves freely under gravity. After 5 seconds the components of velocity of P are $20\,\text{m s}^{-1}$ horizontally and $14\,\text{m s}^{-1}$ vertically *downwards*.

(a) Show that $\tan\alpha = \frac{7}{4}$.

(b) Calculate, to 1 decimal place, the value of V.

(c) Find the greatest height above the ground reached by P.

(d) Calculate the speed of P, in m s^{-1} to 1 decimal place, 7 seconds after leaving O. [E]

9

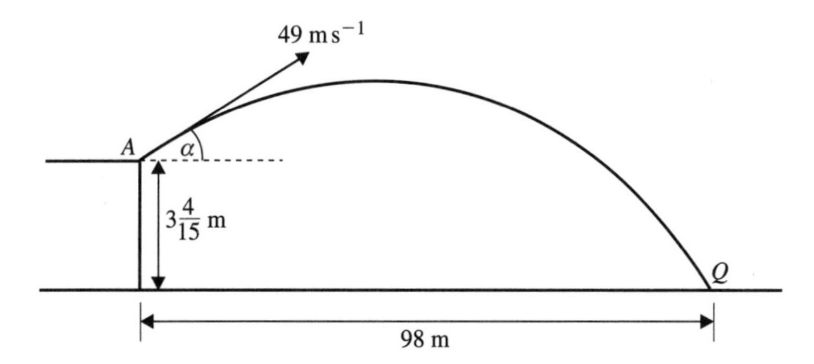

A golf ball is projected with speed $49\,\text{m}\,\text{s}^{-1}$ at an angle of elevation α from a point A on the first floor of a golf driving range. Point A is at a height of $3\frac{4}{15}$ m above horizontal ground. The ball first strikes the ground at a point Q which is at a horizontal distance of 98 m from the point A as shown in the diagram.

(a) Show that

$$6\tan^2\alpha - 30\tan\alpha + 5 = 0$$

(b) Hence find, to the nearest degree, the two possible angles of elevation.

(c) Find, to the nearest second, the smallest possible time of direct flight from A to Q. [E]

10 A particle P moves in a straight line in such a way that, at time t seconds, its velocity $v\,\text{m}\,\text{s}^{-1}$ is given by

$$v = 12t - 3t^2, \quad 0 \leqslant t \leqslant 5$$
$$v = -375t^{-2}, \quad t > 5$$

When $t = 0$, P is at the point O.
Calculate the displacement of P from O

(a) when $t = 5$ (b) when $t = 6$. [E]

11 A particle P moves from rest, at a point O at time $t = 0$ seconds, along a straight line. At any subsequent time t seconds the acceleration of P is proportional to $(7 - t^2)\,\text{m}\,\text{s}^{-2}$ and the displacement of P from O is s metres. The speed of P is $6\,\text{m}\,\text{s}^{-1}$ when $t = 3$.

(a) Show that:

$$s = \tfrac{1}{24}t^2(42 - t^2)$$

(b) Find the total distance, in metres, that P moves before returning to O. [E]

12 At time t seconds, where $t \geqslant 0$, the velocity $v\,\text{m s}^{-1}$ of a particle P moving in a straight line is given by

$$v = 4t - t^2$$

When $t = 0$, P is at the point O.

(a) Find the acceleration of P at time t.

(b) Calculate the time at which P returns to O. [E]

13 A particle P moves along the x-axis. It passes through the origin O at time $t = 0$ with speed $7\,\text{m s}^{-1}$ in the direction of x increasing.

At time t seconds the acceleration of P in the direction of x increasing is $(20 - 6t)\,\text{m s}^{-2}$.

(a) Show the velocity $v\,\text{m s}^{-1}$ of P at time t seconds is given by

$$v = 7 + 20t - 3t^2$$

(b) Show that $v = 0$ when $t = 7$ and find the greatest speed of P in the interval $0 \leqslant t \leqslant 7$.

(c) Find the distance travelled by P in the interval $0 \leqslant t \leqslant 7$. [E]

14 A particle P moves such that, at time t seconds, its position vector $\mathbf{r}$ metres relative to a fixed origin O is given by

$$\mathbf{r} = ct^2\mathbf{i} + (t^3 - 4t)\mathbf{j}$$

where c is a positive constant. When $t = 2$, the speed of P is $10\,\text{m s}^{-1}$.

(a) Find the value of c.

(b) Find the acceleration vector of P when $t = 2$. [E]

15 A particle P moves so that at time t seconds its position vector, $\mathbf{r}$ metres, relative to a fixed origin O, is given by:

$$\mathbf{r} = (2t - t^3 - 1)\mathbf{i} + (2t^2 - 2)\mathbf{j}$$

(a) Find the velocity of P when $t = 2$.

(b) Hence find the speed of P when $t = 2$. [E]

16 With respect to a fixed origin O, the position vector, $\mathbf{r}$ metres, of a particle P, of mass 0.15 kg, at time t seconds is given by

$$\mathbf{r} = t^3\mathbf{i} - 4t^2\mathbf{j},\ t \geqslant 0$$

(a) Find, in m s^{-1}, the speed of P when $t = 2$.

(b) Find, in N to 2 decimal places, the magnitude of the resultant force acting on P when $t = 2$. [E]

17 A particle moves so that at time t seconds its position vector, $\mathbf{r}$ m, relative to a fixed origin is given by:

$$\mathbf{r} = (t^2 - 4t)\mathbf{i} + (t^3 + at^2)\mathbf{j}$$

where a is a constant.

(a) Find an expression for the velocity of the particle at time t seconds.

(b) Given that the particle comes to instantaneous rest, find the value of a. [E]

18 Two particles, A and B, move in the plane of cartesian coordinate axes Ox, Oy. At time t seconds the position vectors of A and B, referred to the origin O, are $(t^2\mathbf{i} + 4t\mathbf{j})$ m and $[2t\mathbf{i} + (t+1)\mathbf{j}]$ m respectively.

(a) Prove that the particles never collide.

(b) Show that the velocity of B is constant, and calculate the magnitude and direction of this velocity.

(c) Find the value of t when A and B have parallel velocities, and find the distance between A and B at this instant. [E]

19 A particle X, moving along a straight line with constant speed $4\,\text{m s}^{-1}$, passes through a fixed point O. Two seconds later another particle Y, moving along the same straight line and in the same direction, passes through O with speed $6\,\text{m s}^{-1}$. Given that Y is subject to a constant deceleration of magnitude $2\,\text{m s}^{-2}$:

(a) state the velocity and displacement of each particle t seconds after Y passed through O

(b) find the shortest distance between the particles after they have both passed through O

(c) find the value of t when the distance between the particles has increased to 23 m. [E]

20 Particles of mass 2 kg, 3 kg and p kg are placed at points whose coordinates are (1, 3), (4, 6) and (7, −8) respectively. Given that the centre of mass of these particles lies on the x-axis, find p. [E]

21 A uniform wire of length $12a$ is bent to form the sides of a triangle ABC in which $AB = 3a$, $BC = 4a$ and angle ABC is $90°$. Find the distances of the centre of mass of the triangular wire ABC from AB and BC. [E]

22 In $\triangle ABC$, $AB = 5a$, $BC = 12a$ and $AC = 13a$. Particles of mass m, $2m$ and $3m$ are placed at the vertices A, B and C respectively. Calculate, in terms of a, the distance of the centre of mass of the three particles from (a) side AB, (b) side BC. [E]

23 A uniform rectangular lamina $ABCD$ is of mass $3M$; $AB = DC = 4$ cm and $BC = AD = 6$ cm. Particles, each of mass M, are attached to the lamina at B, C and D. Calculate the distance of the centre of mass of the loaded lamina

(a) from AB (b) from BC. [E]

24 Three particles of masses 0.1 kg, 0.2 kg and 0.3 kg are placed at the points with position vectors $(2\mathbf{i} - \mathbf{j})$ m, $(2\mathbf{i} + 5\mathbf{j})$ m and $(4\mathbf{i} + 2\mathbf{j})$ m respectively. Find the position vector of the centre of mass of these particles. [E]

25 (i) A uniform lamina $ABCEF$ is obtained from a rectangle $ABCD$, with $AB = CD = 8$ cm and $BC = AD = 6$ cm, by removing the $\triangle EDF$, where E, F lie on CD, AD respectively, with $CE = 2$ cm and $AF = 3$ cm.

(a) Find the distances of the centre of mass of the lamina $ABCEF$ from AB and AD.

(b) The lamina is suspended freely from F and hangs in equilibrium under gravity. Find the angle which AF makes with the vertical. [E]

(ii) Explain what is meant by the word 'lamina'. The lamina in part (i) is described as being 'suspended freely from F'. What assumptions does this allow you to make about the forces acting on the lamina when it is suspended from F?

26 The figure shows a uniform lamina $ABCDE$; $ABCE$ is a rectangle in which $AB = 7$ cm and $BC = 12$ cm. The triangle ECD is isosceles with D distant 12 cm from EC. Calculate the distance of the centre of mass of the lamina from AB. [E]

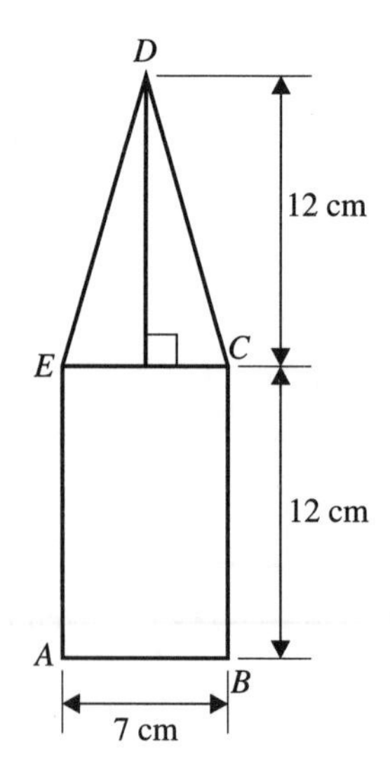

27 A thin uniform triangular plate ABC has $AB = AC = 5$ cm and $BC = 8$ cm. The vertex B is smoothly hinged to a fixed point and the plate hangs in equilibrium in a vertical plane. Calculate, to the nearest degree, the angle made by BC with the vertical. [E]

28 A uniform rectangular plate $ABCD$ with $AB = 6a$ and $BC = 4a$ is of mass 6 kg. Points H and E are on the sides BA and BC respectively such that $BH = BE = 2a$. The square $BEFH$ is removed from the plate.

(a) Show that the centre of mass of the remaining plate $AHFECD$ is at a distance:

(i) $2\frac{1}{5}a$ from AB

(ii) $2\frac{3}{5}a$ from AD.

The plate $AHFECD$ is now freely suspended from the point A.

(b) Find, to the nearest degree, the angle between AD and the downward vertical. [E]

29 A uniform rectangular plate $OABC$ has mass $4m$, $OA = BC = 2d$ and $OC = AB = d$. Particles of mass $2m$, m and $3m$ are attached at A, B and C respectively on the plate. Find, in terms of d, the distance of the centre of mass of the loaded plate:

(a) from OA

(b) from OC.

The corner O of the loaded plate is freely hinged to a fixed point and the plate hangs at rest in equilibrium.

(c) Calculate, to the nearest degree, the angle between OC and the downward vertical. [E]

30

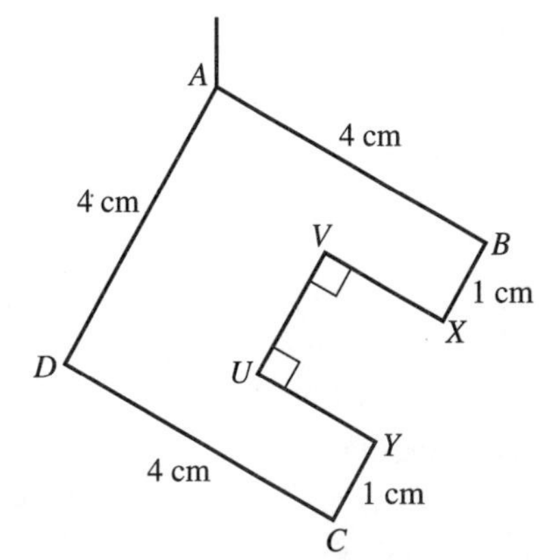

The diagram shows an ear-ring made from a uniform square lamina $ABCD$, which has each side of length 4 cm. Points X and Y are on the side BC and such that $BX = CY = 1$ cm. The square portion $XYUV$ is removed and the resulting ear-ring is suspended from the corner A. The ear-ring hangs in equilibrium.

The centre of mass of this ear-ring is G.

(a) State the distance, in cm, of G from AB.

(b) Find the distance, in cm, of G from AD.

(c) Find, to the nearest degree, the acute angle made by AD with the downward vertical. [E]

31 (i) The unit vectors **i** and **j** are horizontal and vertically upwards respectively. A particle P is projected with velocity $(14\mathbf{i} + 21\mathbf{j})\,\text{m s}^{-1}$ from a point O on horizontal ground and moves freely under gravity. At time T seconds after projection, P strikes the ground at the point C. Calculate:

(a) the value of T

(b) the distance OC

(c) the position vector of H, the point at which P achieves its greatest height above the ground.

Use the principle of conservation of energy to show that at an instant when P is at height 10 m above the ground the speed of P is $21\,\text{m s}^{-1}$. [E]

(ii) Write down two assumptions which you have made about the forces on the particle during its flight.

32 A particle of mass 0.4 kg is projected vertically upwards from a point A and moves freely under gravity. Given that the kinetic energy of the particle at height 5 m above A is 2.45 J, find the speed of projection of the particle. [E]

33 The total mass of a cyclist and his machine is 140 kg. The cyclist rides along a horizontal road against a constant total resistance of magnitude 50 N. Find, in J, the total work done in increasing his speed from $6\,\text{m s}^{-1}$ to $9\,\text{m s}^{-1}$ whilst travelling a distance of 30 m. [E]

34 (a) A car of mass 750 kg, moving along a level road, has its speed reduced from $25\,\text{m s}^{-1}$ to $15\,\text{m s}^{-1}$ by the brakes which produce a constant retarding force of 2250 N. Calculate the distance covered whilst the speed is being reduced.

(b) Write down the assumptions you have made about the car in part (a) and the forces on it during its journey. [E]

35 A ball is projected from a point A on horizontal ground, with speed $14\,\text{m s}^{-1}$ at an angle of elevation $\alpha°$, and moves freely under gravity. At the instant when the ball is at a point P, which is 5 m above the ground, the components of velocity of the ball horizontally and vertically *upwards* are both $u\,\text{m s}^{-1}$.

(a) By using energy considerations, or otherwise, show that $u = 7$.

(b) Obtain the value of α.

(c) Find, in seconds to 2 decimal places, the time taken by the ball to reach its greatest height above the ground.

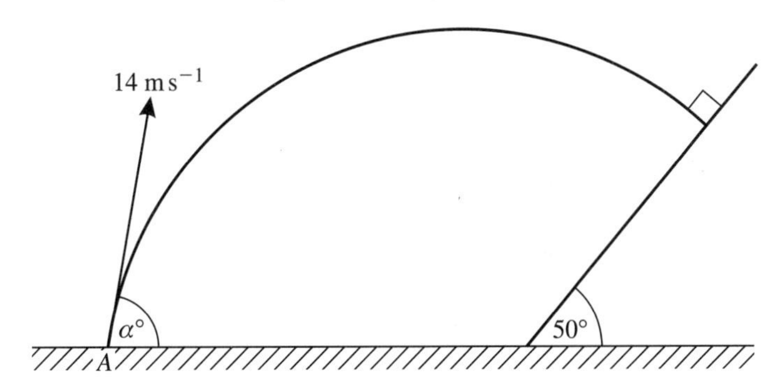

The diagram shows a fixed plane inclined at an angle of 50° to the horizontal. The ball strikes this plane, at right angles, with speed V m s^{-1}.

(d) Calculate, to 1 decimal place, the value of V. [E]

36 The unit vectors **i** and **j** are parallel to the coordinate axes Ox and Oy respectively. Two particles, P and Q, each of mass 0.4 kg, leave the origin O simultaneously and move in the plane of Ox and Oy. The velocity vectors of P and Q at time t seconds after leaving O are $(4t\mathbf{i} + 6t\mathbf{j})$ m s^{-1} and $(3t^2\mathbf{i} - 2\mathbf{j})$ m s^{-1} respectively.

(a) Calculate, in J, the kinetic energy of Q at the instant when $t = 2$.

(b) Find the acceleration vector of P and hence prove that the force **F** acting on P is constant.

(c) Calculate, to the nearest 0.1 N, the magnitude of **F** and find, to the nearest degree, the angle between Oy and the line of action of **F**.

(d) Find the distance between P and Q at the instant when $t = 2$. [E]

37 A rock-climber of mass 60 kg does 6468 J of work in slowly climbing up a fixed vertical rope.

(a) Calculate the height through which he climbs.

(b) Given that his climb takes 40 s, calculate, to the nearest watt, his average rate of working. [E]

38 A lorry is moving along a straight level road at constant speed 48 km h^{-1} against resistances of total magnitude of 900 N. Find, in kW, the rate at which the engine of the lorry is working. [E]

39 The total mass of a man and his bicycle is 80 kg. The man freewheels down a slope inclined at an angle θ to the horizontal, where $\sin\theta = \frac{1}{7}$, with constant acceleration of magnitude $0.9\,\text{m s}^{-2}$.

(a) Prove that the total magnitude of the resistive forces opposing the motion is 40 N.

(b) Find the time required for the man to cover 180 m from rest.

The man cycles up the same slope at constant speed $6\,\text{m s}^{-1}$, the resistive forces remaining unchanged.

(c) Find, in W, the power that must be exerted by the man.

(d) If the man suddenly increases his work rate by 240 W, find the magnitude of his acceleration up the slope at this instant. [E]

40 The non-gravitational resistive forces opposing the motion of a lorry of mass 2200 kg are constant and total 3100 N.

(a) The lorry is moving along a straight horizontal road at a constant speed of $20\,\text{m s}^{-1}$. Calculate, in kW, the rate at which the engine of the lorry is working.

(b) If this rate of working is suddenly decreased by 12 kW find, in m s^{-2}, the immediate retardation of the lorry.

(c) The lorry moves up a hill inclined at an angle θ to the horizontal, where $\sin\theta = \frac{10}{49}$, with the engine working at 90 kW. Calculate, in m s^{-1}, the greatest steady speed which the lorry can maintain up this hill. [E]

41 An engine, of mass 25 tonnes, pulls a truck of mass 10 tonnes along a railway line (1 tonne = 1000 kg). The frictional resistances to the motion of the engine and truck are constant and of magnitude 50 N per tonne mass. When the train travels horizontally the tractive force exerted by the engine is 26 kN. Calculate:

(a) the acceleration, in m s^{-2}, of the engine and truck

(b) the tension, in kN, in the coupling between the engine and the truck.

The engine and truck now start to climb a slope whose inclination to the horizontal is α, where $\sin\alpha = \frac{1}{70}$, and the frictional resistances are unaltered. At a certain instant the engine and truck are moving up the slope with speed $10\,\text{m s}^{-1}$ and acceleration $0.6\,\text{m s}^{-2}$. Calculate, at that instant:

(c) the tractive force, in kN, exerted by the engine

(d) the power, in kW, developed by the engine. [E]

42 The engine of a car is working at a constant rate of 6 kW in driving the car along a straight horizontal road at a constant speed of $54\,\text{km}\,\text{h}^{-1}$. Find, in N, the resistance to the motion of the car. [E]

43 The force $(6\mathbf{i} + 4\mathbf{j})$ newtons acts on a particle of mass 2 kg for 7 seconds. The particle starts from rest and at the end of the 7 seconds its velocity is $(p\mathbf{i} + q\mathbf{j})\,\text{m}\,\text{s}^{-1}$. Find the values of p and q. [E]

44 The mass of a car is 800 kg. Its engine works at a constant rate of 20 kW, and the motion of the car is subject to a constant resistance of magnitude 600 N.

(a) Find, in kJ to 2 significant figures, the kinetic energy of the car when it is travelling at maximum speed on a level road.

(b) Find, in $\text{m}\,\text{s}^{-2}$ the acceleration of the car when it is travelling at $25\,\text{m}\,\text{s}^{-1}$ on a level road.

The car now ascends a straight road, inclined at $5°$ to the horizontal, with the same power output and against the same constant resistance.

(c) Find, in $\text{m}\,\text{s}^{-1}$ to 2 significant figures, the maximum speed of the car up the slope. [E]

Collisions

4

4.1 Collisions in the real world

Collisions between two freely moving bodies or between a freely moving body and a fixed object are everyday occurrences. In general as a result of a collision the bodies bounce away from each other. Consider the collision of two tennis balls. The time during which they are in contact may be divided into two parts:

(1) the period of compression,

(2) the period of restitution, during which they recover their shape.

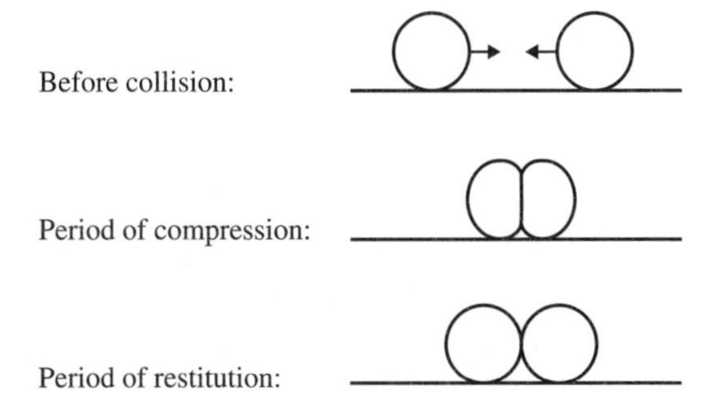

The property which causes bodies like tennis balls to recover their shape and hence causes them to rebound is called **elasticity**.

In dealing with the impact of elastic bodies it is usually assumed that they are **smooth**. This means that the mutual reaction acts along the common normal at the point of impact. When two elastic spheres collide the mutual reaction acts along their line of centres as this is the common normal. The bodies are said to collide **directly** if the motion of each is along this common normal. For example two spheres of the same radius collide directly if they are moving parallel to their line of centres.

In this chapter we will consider the collision of 'elastic particles' moving on a straight line. Many collisions in the real world may be modelled by this situation by making appropriate assumptions, for example the collision of two snooker balls.

4.2 Collisions of particles

In Chapter 5 of Book M1 the collision of two elastic particles was considered. Remember that:

(1) The momentum of a particle of mass m moving with velocity $\mathbf{v}$ is $m\mathbf{v}$. As $\mathbf{v}$ is a vector it follows that **momentum is also a vector**.

(2) **The impulse of a force = change in momentum produced**

$$\mathbf{I} = m\mathbf{v} - m\mathbf{u}$$

This is sometimes called the **impulse–momentum principle**.

(3) The **principle of conservation of momentum**:

■ **Total momentum before collision = total momentum after collision**

For particles moving along a straight line this gives:

Before collision: $\rightarrow u_1$ ○ m_1 $\rightarrow u_2$ ○ m_2

After collision: $\rightarrow v_1$ ○ m_1 $\rightarrow v_2$ ○ m_2

$$m_1u_1 + m_2u_2 = m_1v_1 + m_2v_2$$

Example 1

A particle of mass 5 kg is moving with velocity $(3\mathbf{i} + 4\mathbf{j})\,\text{m s}^{-1}$ when it is given an impulse $(-2\mathbf{i} + 6\mathbf{j})$ N s. Find the velocity of the particle after the impact.

If the velocity of the particle after impact is $\mathbf{v}\,\text{m s}^{-1}$ then the change in momentum is:

$$5[\mathbf{v} - (3\mathbf{i} + 4\mathbf{j})]\,\text{kg m s}^{-1}$$

From the impulse–momentum principle this is equal to the impulse.

So:

$$5\mathbf{v} - 5(3\mathbf{i} + 4\mathbf{j}) = -2\mathbf{i} + 6\mathbf{j}$$

$$5\mathbf{v} = -2\mathbf{i} + 6\mathbf{j} + 15\mathbf{i} + 20\mathbf{j}$$

$$= 13\mathbf{i} + 26\mathbf{j}$$

and:

$$\mathbf{v} = \frac{1}{5}(13\mathbf{i} + 26\mathbf{j})$$

The velocity of the particle after impact is $\frac{1}{5}(13\mathbf{i} + 26\mathbf{j})\,\text{m s}^{-1}$.

Example 2

A particle of mass 0.5 kg is moving horizontally in a straight line with a speed of $20\,\text{m s}^{-1}$. It is hit by a bat, and moves back along its original path with a speed of $25\,\text{m s}^{-1}$. Find the magnitude of the impulse exerted on the particle by the bat. Describe a 'realistic situation' this could be used to model.

The motion in this problem is along a straight line. The vector properties of the momentum and the impulse may be taken into account by taking the direction of the impulse as positive.

Before impact: Bat $\rightarrow I$ $\leftarrow 20\,\text{m s}^{-1}$

After impact: $\rightarrow 25\,\text{m s}^{-1}$

Therefore:

$$\text{change in momentum is} = 0.5(25 - (-20))$$
$$= 22.5$$

This is equal to the impulse, so:

$$\text{I} = 22.5\,\text{N s}$$

The magnitude of the impulse exerted by the bat on the particle is 22.5 N s.

This could be used to model a cricket bat striking a cricket ball. The ball will then be modelled by a particle. This is a reasonable assumption as the dimensions of the ball are small compared to those of the bat.

Example 3

Two particles P and Q of mass 2 kg and 3 kg respectively are moving towards each other along the same straight line with speeds $4\,\text{m s}^{-1}$ and $5\,\text{m s}^{-1}$ respectively. After the collision the particles coalesce, that is move as a single body. Find the speed of the combined particle after the collision. Describe a situation this could be used to model.

It is important to remember that momentum is a vector. The equation:

total momentum before collision = total momentum after collision

is therefore a **vector equation**.

As in the previous example it is necessary to choose a particular direction as positive. It is immaterial which direction is chosen. Choose the direction of motion of P as positive and let the speed of the combined particle after impact be $v\,\text{m s}^{-1}$ in that direction. The figure shows the situation before and after impact.

Before impact: P (2 kg) moving at $4\,\text{m s}^{-1}$ →; Q (3 kg) moving at $5\,\text{m s}^{-1}$ ←

After impact: $P + Q$ (5 kg) moving at $v\,\text{m s}^{-1}$ →

The momentum before impact $= 2 \times 4 - 3 \times 5$

The momentum after impact $= 5v$

By the principle of conservation of momentum:

$$2 \times 4 - 3 \times 5 = 5v$$

So:

$$v = \frac{-7}{5} = -1.4$$

The speed of the combined particle is $1.4\,\text{m s}^{-1}$. Since v is negative, the speed is in the opposite direction to the original motion of particle P.

This could be used to model two balls of putty which will stick together on impact.

It could also be used to model two small trucks, moving along a track, which couple when they collide.

Exercise 4A

1 A particle of mass 0.5 kg is moving with velocity $(6\mathbf{i} + 2\mathbf{j})\,\text{m s}^{-1}$ when it is given an impulse $(3\mathbf{i} - 2\mathbf{j})\,\text{N s}$. Find the new velocity of the particle. Describe a 'realistic situation' this could model.

2 A particle of mass 0.5 kg is moving with velocity $(4\mathbf{i} + 7\mathbf{j})\,\text{m s}^{-1}$ when it strikes a fixed wall. It rebounds with velocity $(2\mathbf{i} + 3\mathbf{j})\,\text{m s}^{-1}$. Find the impulse exerted on the particle by the wall.

3 An impulse $(4\mathbf{i} - 8\mathbf{j})\,\text{N s}$ is applied to a particle of mass 2 kg. The velocity of the mass after the impact is $(3\mathbf{i} + 5\mathbf{j})\,\text{m s}^{-1}$. Find its velocity before impact.

4 A particle of mass 1.5 kg is moving along a straight line with speed $12\,\text{m s}^{-1}$. It receives an impulse I N s and continues to move in the same direction but with a speed of $24\,\text{m s}^{-1}$. Find I.

5 A particle of mass 3 kg is moving along a straight line with speed $8\,\text{m s}^{-1}$. It receives an impulse I N s and then moves along the same straight line but in the opposite direction, with a speed of $4\,\text{m s}^{-1}$. Find I.

6 A particle of mass 2 kg is falling at a speed of $12\,\text{m s}^{-1}$ when it strikes the ground. The particle is brought to rest by the impact. Find the impulse exerted by the ground on the particle. Describe a 'realistic situation' this could model.

7 A particle of mass 3 kg falls from rest at a height of 5 m above a horizontal plane. It rebounds to a height of 3 m. Find the impulse exerted on the particle by the plane.

8 Two particles A and B of mass 3 kg and 4 kg respectively are moving towards each other along the same straight line with speeds $5\,\text{m s}^{-1}$ and $3\,\text{m s}^{-1}$ respectively. After the collision the particles coalesce. Find the magnitude and direction of the velocity of the combined particle after the impact. What realistic situation could this model?

9 Two particles P and Q of mass 4 kg and 2 kg respectively travel towards each other along the x-axis. The velocities of P and Q are $6\mathbf{i}\,\text{m s}^{-1}$ and $(-2\mathbf{i})\,\text{m s}^{-1}$ respectively. The particles collide and after the collision Q has a velocity $3\mathbf{i}\,\text{m s}^{-1}$. Find the velocity of P after the collision.

10 Two particles A and B of mass 3 kg and 5 kg respectively move along the same straight line and collide. Just before the collision A is moving with speed $8\,\text{m s}^{-1}$ and B is moving with speed $4\,\text{m s}^{-1}$ in the opposite direction. After the collision the particle A rebounds with speed $7\,\text{m s}^{-1}$. Find the velocity of B after the collision.

4.3 Newton's law of restitution for direct impact

A **direct** impact is a collision between two particles which are moving along the same straight line.

When a ball strikes a fixed wall the speed with which it rebounds depends on the material of which the ball is made. In a similar way when two elastic particles collide their speeds after the collision depend on the material they are made from.

Newton's law of restitution, sometimes called Newton's experimental law, defines precisely how the speeds of the particles after the collision depend on the nature of the particles as well as on the speeds before the collision. It states that:

■ $$\frac{\textbf{speed of separation of particles}}{\textbf{speed of approach of particles}} = e$$

Notice that both the numerator and the denominator are positive in the statement.

The constant e is called the **coefficient of restitution**.

This law only holds when the collision takes place in free space or on a **smooth** surface. (On a **rough** surface the speed of separation of two particles may be affected by 'spin'.) The value of e satisfies the inequality

$$0 \leqslant e \leqslant 1$$

For two perfectly elastic particles $e = 1$.

For two inelastic particles, that is two particles which coalesce on impact, $e = 0$.

For a pair of glass marbles $e = 0.95$.

For a pair of billiard balls $e = 0.8$.

For a pair of lead spheres $e = 0.2$.

It is important to be clear what 'the speed of separation' and 'the speed of approach' are in any given situation. The diagrams below illustrate three typical situations.

(1) Before impact: $\rightarrow u_1$ ○ $\rightarrow$ 0 (at rest) ○

Speed of approach $= u_1 - 0 = u_1$

After impact: $\rightarrow$ 0 (at rest) ○ $\rightarrow v_2$ ○

Speed of separation $= v_2 - 0 = v_2$

(2) Before impact: $\rightarrow u_1$ ○ $\rightarrow u_2$ ○

Speed of approach $= u_1 - u_2$

(u_1 must be greater than u_2 or there would be no collision.)

After impact: $\rightarrow v_1$ $\rightarrow v_2$

Speed of separation $= v_2 - v_1$

(v_2 must be greater than v_1 or they would not separate.)

(3) Before impact: $\rightarrow u_1$ $u_2 \leftarrow$

Speed of approach $= u_1 + u_2$

After impact: $v_1 \leftarrow$ $\rightarrow v_2$

Speed of separation $= v_1 + v_2$

Example 4

Find the value of e in the situations shown.

(a) Before impact: $\rightarrow 6\,\text{m s}^{-1}$ $\rightarrow 0$ (at rest)

After impact: $\rightarrow 2\,\text{m s}^{-1}$ $\rightarrow 4\,\text{m s}^{-1}$

(b) Before impact: $\rightarrow 4\,\text{m s}^{-1}$ $2\,\text{m s}^{-1} \leftarrow$

After impact: $1\,\text{m s}^{-1} \leftarrow$ $\rightarrow 3\,\text{m s}^{-1}$

(a)

$$e = \frac{\text{speed of separation}}{\text{speed of approach}}$$

$$= \frac{4-2}{6-0} = \frac{2}{6} = \frac{1}{3}$$

(b)

$$e = \frac{\text{speed of separation}}{\text{speed of approach}}$$

$$= \frac{3+1}{4+2} = \frac{4}{6} = \frac{2}{3}$$

Example 5

Find the value of v in the situation shown, given that the coefficient of restitution e is $\frac{2}{3}$.

Before impact: $6\,\text{m s}^{-1} \leftarrow$ $3\,\text{m s}^{-1} \leftarrow$

After impact: $\rightarrow v\,\text{m s}^{-1}$ $\rightarrow 7\,\text{m s}^{-1}$

As: $$e = \frac{\text{speed of separation}}{\text{speed of approach}}$$

so: $$\frac{2}{3} = \frac{7 - v}{6 + 3}$$

and: $$2(9) = 3(7 - v)$$

Therefore: $$v = 1$$

Problems involving the direct collision of two particles may be solved by writing down two equations:

(1) the conservation of linear momentum,

(2) Newton's law of restitution.

These two simultaneous equations may be solved to find two unknowns.

Example 6

Two particles A and B of mass 0.2 kg and 0.5 kg respectively are moving towards each other along the same straight line on a smooth horizontal table. Particle A has speed $12\,\text{m s}^{-1}$ and particle B has speed $2\,\text{m s}^{-1}$. Given that the coefficient of restitution between the particles is $\frac{1}{2}$, find:

(a) the speeds of A and B after the impact,

(b) the magnitude of the impulse given to each particle.

Describe a realistic situation this could be used to model.

(a) First draw diagrams of the situations before and after impact.

Before impact: $\rightarrow 12\,\text{m s}^{-1}$ A (0.2 kg); $2\,\text{m s}^{-1} \leftarrow$ B (0.5 kg)

After impact: $\rightarrow v_1\,\text{m s}^{-1}$ A (0.2 kg); $\rightarrow v_2\,\text{m s}^{-1}$ B (0.5 kg)

Using conservation of momentum with $m_1 = 0.2\,\text{kg}$, $m_2 = 0.5\,\text{kg}$, $u_1 = 12\,\text{m s}^{-1}$ and $u_2 = -2\,\text{m s}^{-1}$ gives:

$$(0.2)12 - (0.5)2 = (0.2)v_1 + (0.5)v_2$$

or: $$2v_1 + 5v_2 = 14 \qquad (1)$$

Newton's law of restitution states:

$$e = \frac{\text{speed of separation}}{\text{speed of approach}}$$

So: $$\frac{1}{2} = \frac{v_2 - v_1}{14}$$

and: $$v_2 - v_1 = 7 \qquad (2)$$

Eliminating v_1 between equations (1) and (2) gives:

$$7v_2 = 28$$

So: $$v_2 = 4$$

Substituting this value into equation (2) gives:

$$v_1 = -3$$

The minus sign indicates that A moves in the opposite direction to the arrow in the diagram. The situation after collision is:

$3\ \text{m s}^{-1} \leftarrow$ A (0.2 kg) $\rightarrow 4\ \text{m s}^{-1}$ B (0.5 kg)

In this case the direction of motion of each particle is reversed by the collision.

(b) Recall that 'Impulse = change in momentum' and also that 'momentum is a vector'. Take the initial direction of the velocity of A as positive, that is from left to right.

For particle A we have:

Before impact: $\rightarrow 12\ \text{m s}^{-1}$ (0.2 kg)

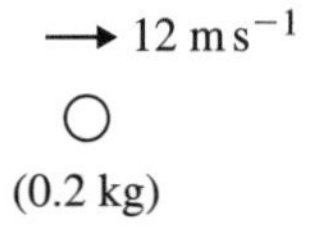

After impact: $3\ \text{m s}^{-1} \leftarrow$ (0.2 kg)

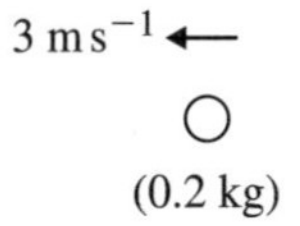

The change in momentum of A is:

$$-(0.2)3 - (0.2)(12) = -3\ \text{N s}$$

So magnitude of impulse on $A = 3\ \text{N s}$

For particle B we have:

Before impact: $2\ \text{m s}^{-1} \leftarrow$ (0.5 kg)

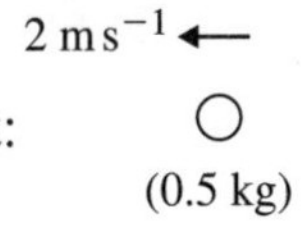

After impact: $\rightarrow 4\ \text{m s}^{-1}$ (0.5 kg)

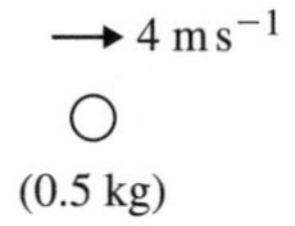

The change in momentum of B is:

$$(0.5)4 - (0.5)(-2) = 3\ \text{N s}$$

So magnitude of impulse on $B = 3\ \text{N s}$.

The two particles A and B may be used to model two small smooth spheres, with the same radius, of mass 0.2 kg and 0.5 kg respectively, moving towards each other on a sheet of ice.

Alternatively A and B may be used to model two toy trucks of mass 0.2 kg and 0.5 kg respectively moving towards each other on a smooth horizontal straight track.

Loss of mechanical energy due to impact

When two elastic particles collide although there is no change in momentum there is in general, ($e \neq 1$), a loss of mechanical energy. When $e = 1$ the particles are perfectly elastic and there is no loss of mechanical energy. (See question 8 in Exercise 4B.) In reality $e \neq 1$ and some of the kinetic energy of the particles is transformed into other forms of energy such as heat energy or sound energy.

For particles moving along a straight line:

Before impact: $\xrightarrow{} u_1$ ○ m_1 $\quad$ $\xrightarrow{} u_2$ ○ m_2

After impact: $\xrightarrow{} v_1$ ○ m_1 $\quad$ $\xrightarrow{} v_2$ ○ m_2

The total kinetic energy before impact is:

$$\tfrac{1}{2}m_1u_1^2 + \tfrac{1}{2}m_2u_2^2$$

The total kinetic energy after impact is:

$$\tfrac{1}{2}m_1v_1^2 + \tfrac{1}{2}m_2v_2^2$$

The loss of mechanical energy due to impact is therefore:

$$\left[\tfrac{1}{2}m_1u_1^2 + \tfrac{1}{2}m_2u_2^2\right] - \left[\tfrac{1}{2}m_1v_1^2 + \tfrac{1}{2}m_2v_2^2\right]$$

For the particles in Example 6 this loss is:

$$\left[\tfrac{1}{2}(0.2)(12)^2 + \tfrac{1}{2}(0.5)(2)^2\right] - \left[\tfrac{1}{2}(0.2)(3)^2 + \tfrac{1}{2}(0.5)(4)^2\right] = 10.5\,\text{J}$$

The change in mechanical energy may also be obtained from:

[change in kinetic energy of particle A] + [change in kinetic energy of particle B]

For the particles in Example 6 this gives:

$$\tfrac{1}{2}(0.2)\left[(12)^2 - (3)^2\right] + \tfrac{1}{2}(0.5)\left[(2)^2 - (4)^2\right] = 10.5\,\text{J}$$

Exercise 4B

1 In each part of this question the two diagrams show the speeds of two particles A and B just before and just after a collision. The particles move on a smooth horizontal plane. Find the coefficient of restitution e in each case.

	Before collision:		After collision:	
	A	B	A	B
(a)	$\rightarrow 12\,\text{m s}^{-1}$	At rest	At rest	$\rightarrow 6\,\text{m s}^{-1}$
(b)	$\rightarrow 6\,\text{m s}^{-1}$	$\rightarrow 2\,\text{m s}^{-1}$	$\rightarrow 3\,\text{m s}^{-1}$	$\rightarrow 4\,\text{m s}^{-1}$
(c)	$\rightarrow 5\,\text{m s}^{-1}$	$4\,\text{m s}^{-1} \leftarrow$	$3\,\text{m s}^{-1} \leftarrow$	$\rightarrow 3\,\text{m s}^{-1}$

2 In each part of this question the two diagrams show the speeds of two particles A and B just before and just after collision on a smooth horizontal plane. The masses of A and B and the coefficient of restitution e are also given. Find the speeds v_1 and v_2 in each case.

	Before collision:		After collision:	
(a) $e = \frac{1}{2}$:	$\rightarrow 8\,\text{m s}^{-1}$ A (0.5 kg)	At rest B (1 kg)	$\rightarrow v_1\,\text{m s}^{-1}$ A (0.5 kg)	$\rightarrow v_2\,\text{m s}^{-1}$ B (1 kg)
(b) $e = \frac{1}{4}$:	$\rightarrow 6\,\text{m s}^{-1}$ A (2 kg)	$\rightarrow 2\,\text{m s}^{-1}$ B (3 kg)	$\rightarrow v_1\,\text{m s}^{-1}$ A (2 kg)	$\rightarrow v_2\,\text{m s}^{-1}$ B (3 kg)
(c) $e = \frac{1}{2}$:	$\rightarrow 4\,\text{m s}^{-1}$ A (4 kg)	$1\,\text{m s}^{-1} \leftarrow$ B (1 kg)	$\rightarrow v_1\,\text{m s}^{-1}$ A (4 kg)	$\rightarrow v_2\,\text{m s}^{-1}$ B (1 kg)
(d) $e = \frac{3}{4}$:	$\rightarrow 3\,\text{m s}^{-1}$ A (1 kg)	$1\,\text{m s}^{-1} \leftarrow$ B (2.5 kg)	$\rightarrow v_1\,\text{m s}^{-1}$ A (1 kg)	$\rightarrow v_2\,\text{m s}^{-1}$ B (2.5 kg)
(e) $e = \frac{1}{5}$:	$\rightarrow 2\,\text{m s}^{-1}$ A (2 kg)	$3\,\text{m s}^{-1} \leftarrow$ B (2 kg)	$\rightarrow v_1\,\text{m s}^{-1}$ A (2 kg)	$\rightarrow v_2\,\text{m s}^{-1}$ B (2 kg)

3 A small smooth sphere A of mass 2 kg is travelling along a straight line on a smooth horizontal plane with speed $6\,\text{m}\,\text{s}^{-1}$ when it collides with a small smooth sphere B of mass 3 kg moving along the same straight line in the same direction with speed $2\,\text{m}\,\text{s}^{-1}$. After the collision A continues to move in the same direction with speed $3\,\text{m}\,\text{s}^{-1}$.

(a) Find the speed of B after the collision.

(b) Find the coefficient of restitution between A and B.

(c) State any assumptions you made in your calculations.

4 Two smooth spheres of equal radius have mass 2.5 kg and 1.5 kg respectively. They are travelling towards each other along the same straight line on a smooth horizontal plane. The heavier sphere has a speed of $3\,\text{m}\,\text{s}^{-1}$ and the other sphere has a speed of $2\,\text{m}\,\text{s}^{-1}$. The spheres collide and after the collision the heavier sphere is at rest.

(a) Find the coefficient of restitution between the spheres.

(b) Find the kinetic energy lost in the collision.

5 A small smooth sphere of mass 3 kg moving on a smooth horizontal plane with a speed of $8\,\text{m}\,\text{s}^{-1}$ collides directly with a sphere of mass 12 kg which is at rest. Given that the spheres move in opposite directions after the collision, obtain the inequality satisfied by e.

6 Two identical smooth spheres each of mass m are projected directly towards each other on a smooth horizontal surface. Each sphere has a speed u and the coefficient of restitution between the spheres is e. Show that the collision between the spheres causes a loss of kinetic energy of $mu^2(1 - e^2)$. Describe a 'realistic situation' this could be used to model. State clearly the assumptions you have made.

7 A sphere of mass m is moving with a speed V along a horizontal straight line. It collides with an identical sphere of mass m moving along the same straight line in the same direction with speed u $(u < V)$. Show that the magnitude of the impulse on either of the spheres is

$$\tfrac{1}{2}m(1+e)(V-u)$$

where e is the coefficient of restitution between the two spheres.

8 A sphere of mass m_1 moving with speed u_1 collides directly with a similar sphere of mass m_2 moving with speed u_2 in the same direction ($u_1 > u_2$). The coefficient of restitution between the two spheres is e. Show that the loss of kinetic energy E due to the collision satisfies the equation

$$2(m_1 + m_2)E = m_1 m_2 (u_1 - u_2)^2 (1 - e^2)$$

4.4 Impact of a particle with a fixed surface

Consider a small sphere, moving on a smooth horizontal surface, which strikes a fixed vertical barrier that is perpendicular to the direction of motion of the sphere as shown in the figure.

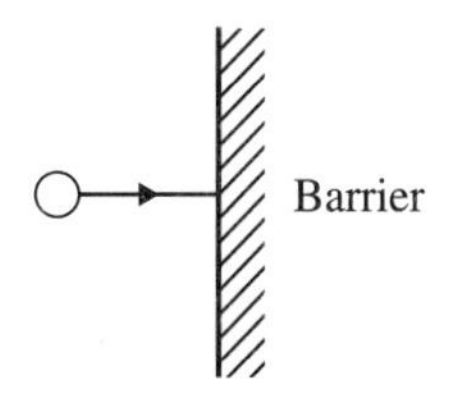

As the sphere is small we can model it by an elastic particle. The phrase commonly used to describe the above situation is 'the particle collides normally with a fixed barrier'.

From your own experience you will know that the direction of motion of the sphere is reversed by the impact. The speed of the sphere after the impact will depend on the material the sphere is made from.

Let the speed before impact be u, the speed after impact be v, and the coefficient of restitution between the particle and the surface be e.

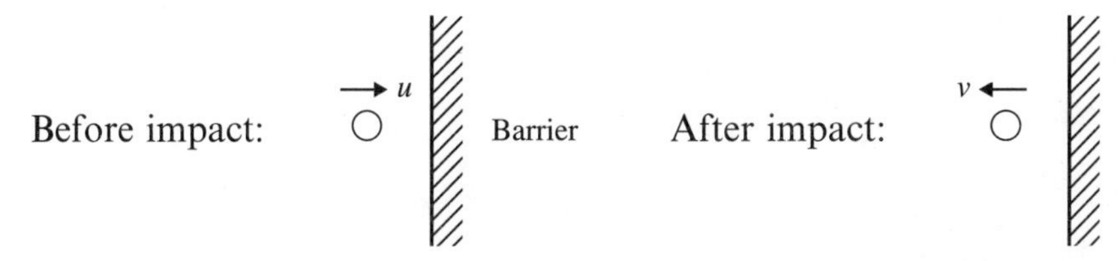

Newton's law of restitution applied to this impact gives:

$$e = \frac{\text{speed of separation}}{\text{speed of approach}}$$

$$= \frac{v - 0}{u - 0} = \frac{v}{u}$$

$$= \frac{\text{speed of rebound}}{\text{speed of approach}}$$

Hence

$$e = \frac{v}{u} \quad \text{or} \quad v = eu$$

So: (speed of rebound) $= e \times$ (speed of approach)

Example 7

A particle is moving horizontally with speed $5\,\text{m}\,\text{s}^{-1}$ when it strikes a vertical wall normally.

It rebounds with a speed of $3\,\text{m}\,\text{s}^{-1}$. Find the coefficient of restitution between the particle and the wall.

Before impact:

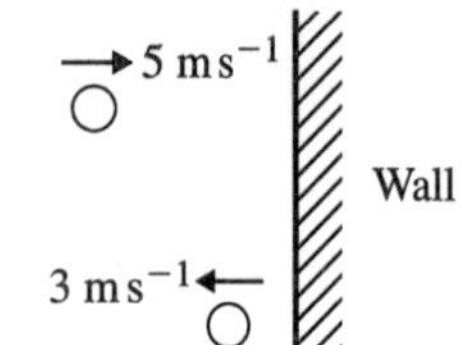

After impact:

By Newton's law of restitution:

$$e = \frac{\text{speed of rebound}}{\text{speed of approach}}$$

$$= \frac{3}{5}$$

The coefficient of restitution is $\frac{3}{5}$.

Example 8

A particle is moving horizontally with speed $3\,\text{m}\,\text{s}^{-1}$ when it strikes a vertical wall normally. The coefficient of restitution between the particle and the wall is $\frac{3}{4}$. Find the speed of rebound.

Before impact:

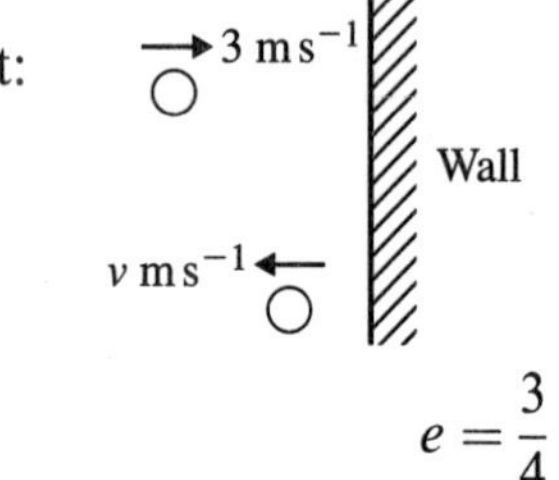

After impact:

$$e = \frac{3}{4}$$

By Newton's law of restitution:

$$e = \frac{\text{speed of rebound}}{\text{speed of approach}}$$

In this case we have:

$$\frac{3}{4} = \frac{v - 0}{3 - 0}$$

So:

$$v = \frac{9}{4}$$

The speed of rebound is $2\frac{1}{4}\,\text{m}\,\text{s}^{-1}$.

Example 9

A small smooth ball falls from rest from a height of 1.35 m above a fixed smooth horizontal plane. It rebounds to a height of 0.6 m. Find the coefficient of restitution between the ball and the plane.

Model the ball by an 'elastic particle'.

The particle moves freely under gravity until it strikes the plane. The speed $v \, \text{m s}^{-1}$ with which it strikes the plane can be found using the constant acceleration formula (Book M1, Chapter 3).

Using $v^2 = u^2 + 2as$ with $u = 0 \, \text{m s}^{-1}$, $s = 1.35 \, \text{m}$ and $a = 9.8 \, \text{m s}^{-2}$ gives:

$$v^2 = 2(9.8)(1.35) \qquad (1)$$

Suppose the particle rebounds from the plane with speed $u \, \text{m s}^{-1}$. Then:

$$e = \frac{u}{v} \text{ or } u = ev \qquad (2)$$

After rebound the particle again moves freely under gravity. Since at the greatest height the speed of the particle is zero, using $v^2 = u^2 + 2as$ with $v = 0 \, \text{m s}^{-1}$, $s = 0.6 \, \text{m}$ and $a = -9.8 \, \text{m s}^{-2}$ gives:

$$u^2 = 2(9.8)(0.6) \qquad (3)$$

Substituting (2) into (3) gives:

$$e^2 v^2 = 2(9.8)(0.6) \qquad (4)$$

and substituting (1) into (4) gives:

$$e^2 2(9.8)(1.35) = 2(9.8)(0.6)$$

or:

$$e^2 = \frac{0.6}{1.35} = \frac{4}{9}$$

So:

$$e = \frac{2}{3} \qquad \text{since } 0 \leqslant e \leqslant 1$$

The coefficient of restitution between the ball and the plane is $\frac{2}{3}$.

Example 10

A particle P falls from rest from a height of 4 m above a smooth horizontal plane. The coefficient of restitution between P and the plane is $\frac{1}{2}$. Find the total distance travelled by P up to the instant when it hits the plane for the 3rd time.

1st rebound

The particle moves freely under gravity until it strikes the plane. The speed with which it strikes the plane may be obtained by using the constant acceleration formula (Book M1, Chapter 3) $v^2 = u^2 + 2as$ with $u = 0\,\text{m s}^{-1}$, $s = 4\,\text{m}$ and $a = 9.8\,\text{m s}^{-2}$.

So: $$v^2 = 2(9.8)4$$

and: $$v = 2(19.6)^{\frac{1}{2}}\,\text{m s}^{-1} \qquad (1)$$

Using Newton's law of restitution the speed of rebound is $v_1 = ev$ where v is given by (1).

So: $$v_1 = \frac{1}{2}2(19.6)^{\frac{1}{2}}\,\text{m s}^{-1} \qquad (2)$$

The height h_1 to which the particle will rise after the first rebound is obtained by using $v^2 = u^2 + 2as$ with $v = 0\,\text{m s}^{-1}$, $u = v_1 = (19.6)^{\frac{1}{2}}\,\text{m s}^{-1}$ and $a = -9.8\,\text{m s}^{-2}$.

So: $$0 = 19.6 - 2(9.8)h_1$$

and: $$h_1 = 1\,\text{m} \qquad (3)$$

The particle rises to a height of 1 m after the 1st rebound. The figure summarises the results so far.

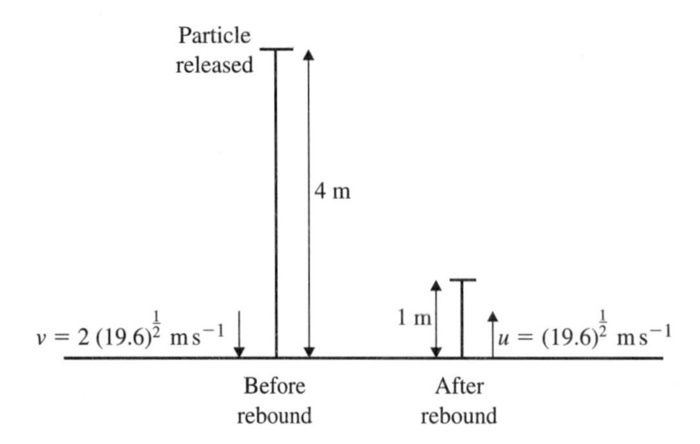

2nd rebound

The particle will now fall freely under gravity and reach the plane for a second time. By symmetry it will reach the plane with a speed of magnitude $v_1 = (19.6)^{\frac{1}{2}}\,\text{m s}^{-1}$ as given by (2).

Using Newton's law of restitution the speed of the second rebound v_2 is:

$$v_2 = ev_1$$

So: $$v_2 = \frac{1}{2}(19.6)^{\frac{1}{2}}\,\text{m s}^{-1} \qquad (4)$$

The height h_2 to which the particle will rise after the second rebound is obtained as before by using $v^2 = u^2 + 2as$ but now with $v = 0\,\text{m s}^{-1}$, $u = v_2 = \frac{1}{2}(19.6)^{\frac{1}{2}}\,\text{m s}^{-1}$ and $a = -9.8\,\text{m s}^{-2}$.

So:
$$0 = \frac{19.6}{4} - 2(9.8)h_2$$

and:
$$h_2 = \frac{1}{4}\text{ m} = 0.25\text{m} \qquad (5)$$

The particle rises to a height of 0.25 m after the second rebound. The figure summarises the results for the second rebound.

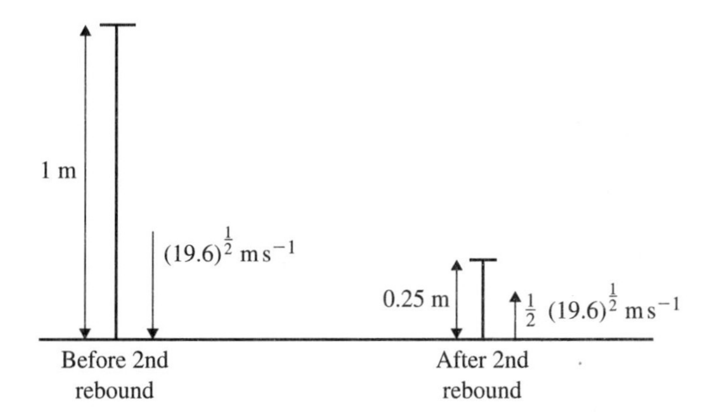

The particle will now fall freely under gravity a distance of 0.25 m and reach the plane for a third time.

The total distance travelled by the particle up to the instant when it hits the plane for the third time is:

(distance travelled before 1st rebound)

+ (distance travelled between 1st and 2nd rebounds)

+ (distance travelled between 2nd and 3rd rebounds)

Using (3) and (5) gives:

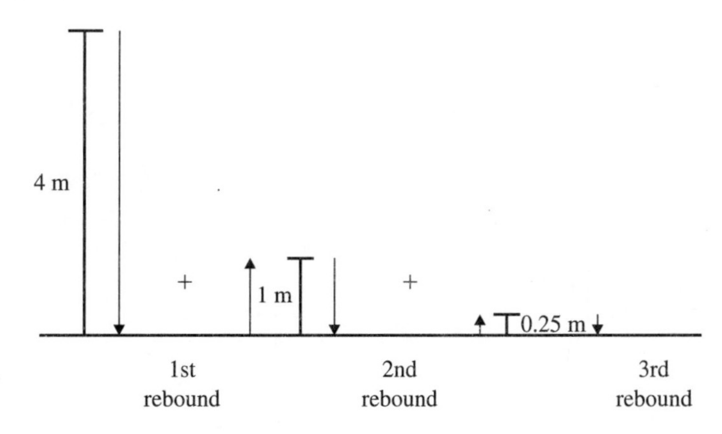

Hence total distance is:

$$4 + (2 \times 1) + (2 \times 0.25) = 4 + 2 + 0.5 = 6.5\text{ m}$$

The total distance travelled by the particle up to the instant when it hits the plane for the 3rd time is 6.5 m.

Exercise 4C

Whenever a numerical value of g is required take $g = 9.8 \text{ m s}^{-2}$

1 A particle collides normally with a fixed vertical wall. The two diagrams show the speeds of the particle before and after collision. In each case find the value of the coefficient of restitution e.

Before impact: After impact:

(a) 8 m s^{-1} Wall 4 m s^{-1} Wall

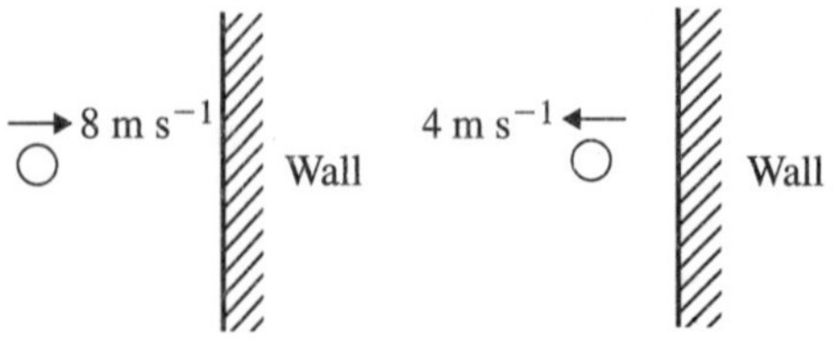

(b) 4 m s^{-1} Wall 3 m s^{-1} Wall

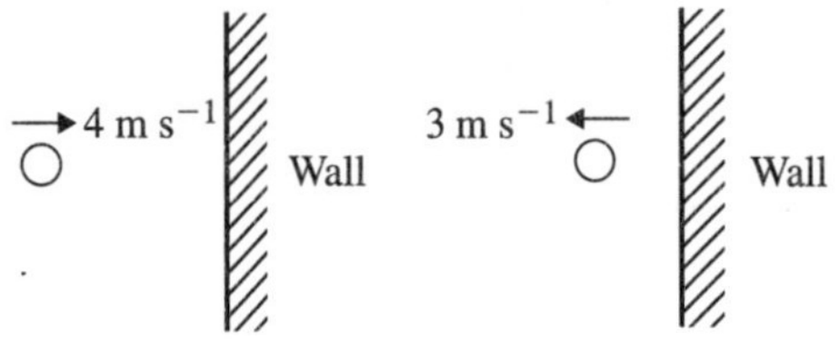

2 A particle collides normally with a fixed vertical wall. The two diagrams show the speed of the particle before impact. In case (a) the coefficient of restitution is $\frac{2}{3}$ and in case (b) the coefficient of restitution is $\frac{1}{4}$. Find the speed of the particle after impact in each case.

Before impact: After impact:

(a) 6 m s^{-1} Wall $v_1 \text{ m s}^{-1}$ Wall

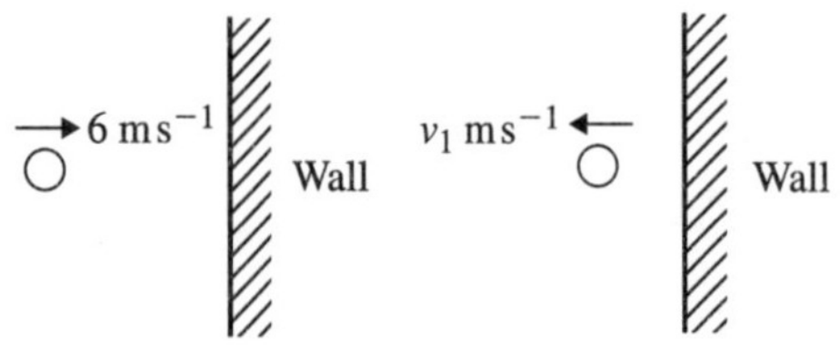

(b) 10 m s^{-1} Wall $v_2 \text{ m s}^{-1}$ Wall

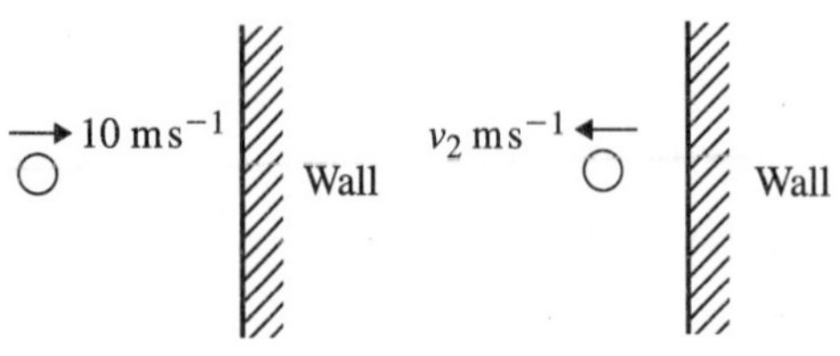

3 A particle collides normally with a fixed vertical wall. The two diagrams show the speed of the particle after impact. In case (a) the coefficient of restitution is $\frac{3}{5}$ and in case (b) the coefficient of restitution is $\frac{1}{2}$. Find u_1 and u_2.

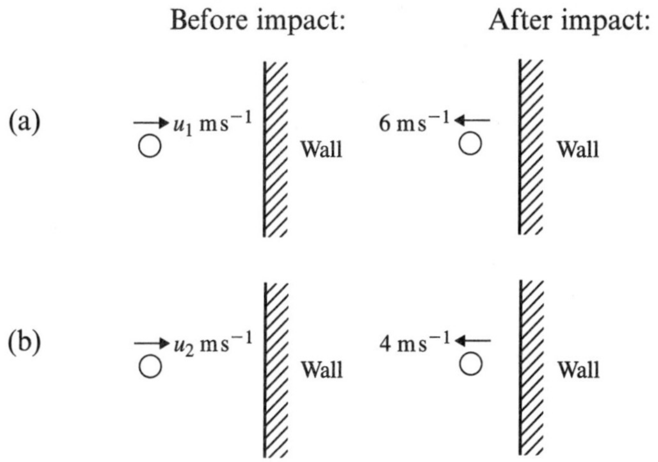

Before impact: After impact:

(a) $u_1\,\text{m s}^{-1}$ Wall $6\,\text{m s}^{-1}$ Wall

(b) $u_2\,\text{m s}^{-1}$ Wall $4\,\text{m s}^{-1}$ Wall

4 A small smooth ball of mass 0.4 kg is moving on a smooth horizontal table with a speed of $12\,\text{m s}^{-1}$ when it collides normally with a fixed smooth vertical wall. It rebounds from the wall with a speed of $9\,\text{m s}^{-1}$.

(a) Find the coefficient of restitution between the ball and the wall.

(b) Find the kinetic energy lost by the ball due to the impact.

(c) Describe a 'realistic situation' this could be used to model.

5 A small smooth ball strikes a smooth vertical wall at right angles. Its kinetic energy after impact is one-half of its initial kinetic energy. Find the coefficient of restitution between the ball and the wall.

6 A particle of mass m is travelling in a straight line with speed u on a smooth horizontal floor. It strikes a fixed smooth vertical wall normally. The kinetic energy lost by the particle due to the collision is E. Show that the coefficient of restitution between the particle and the wall is given by:

$$\sqrt{\left(\frac{mu^2 - 2E}{mu^2}\right)}$$

7 A small ball falls from a height of 4 m on to a smooth horizontal plane. After rebounding from the plane it reaches a height of 1 m. Find the coefficient of restitution between the ball and the plane.

8 A particle falls from rest at a height of h metres on to a smooth horizontal plane. The coefficient of restitution between the particle and the plane is $\frac{1}{5}$. Given that the speed after the second impact is $2\,\text{m}\,\text{s}^{-1}$ find the value of h to the nearest whole number.

9 A small smooth sphere falls from rest on to a smooth horizontal plane. It takes $1\frac{1}{4}$ seconds to reach the plane and another $\frac{3}{4}$ second to come to instantaneous rest. Find the coefficient of restitution between the sphere and the plane. What 'realistic situation' could this model?

10 A small smooth ball bearing falls from rest at a height H above a smooth horizontal plane. The coefficient of restitution between the ball bearing and the plane is e $(e < 1)$. Show that the total distance travelled by the ball bearing before coming to rest is:

$$\left(\frac{1+e^2}{1-e^2}\right)H$$

4.5 Successive impacts

The principles discussed in the previous sections may be applied to situations where three particles collide in pairs or where there are several collisions involving two particles and a wall. In such situations clear diagrams should be drawn showing the 'before' and 'after' information **for each collision**. This is illustrated in the next two examples.

Example 11

Three particles A, B and C have mass 1 kg, 2 kg and 2 kg respectively. They lie in a straight line with B between A and C. The coefficient of restitution between any pair of these spheres is e. Initially B and C are at rest and A is projected towards B with speed $6\,\text{m}\,\text{s}^{-1}$. After collision A rebounds from B with speed $0.5\,\text{m}\,\text{s}^{-1}$.

(a) Show that $e = \frac{5}{8}$.

(b) Find the speed of C after the second impact.

(a) *1st impact*

Before: $\rightarrow 6\,\text{m}\,\text{s}^{-1}$ A (1 kg) $\quad \rightarrow 0\,\text{m}\,\text{s}^{-1}$ B (2 kg)

After: $0.5\,\text{m}\,\text{s}^{-1} \leftarrow$ $\quad \rightarrow u\,\text{m}\,\text{s}^{-1}$

Using the principle of conservation of momentum gives:

$$1 \times 6 + 2 \times 0 = 1 \times (-0.5) + 2u$$

so: $$u = \tfrac{13}{4}$$

This is the speed of B after the 1st impact.

Newton's law of restitution gives:

$$e = \frac{u + 0.5}{6 - 0}$$

And substituting for u gives:

$$e = \frac{\frac{13}{4} + \frac{1}{2}}{6} = \frac{15}{4 \times 6} = \frac{5}{8}$$

(b) *2nd impact*

Before: $\rightarrow \frac{13}{4}\,\mathrm{m\,s^{-1}}$ B (2 kg); $\rightarrow 0\,\mathrm{m\,s^{-1}}$ C (2 kg)

After: $\rightarrow v_1\,\mathrm{m\,s^{-1}}$; $\rightarrow v_2\,\mathrm{m\,s^{-1}}$

Using the principle of conservation of momentum gives:

$$2 \times \frac{13}{4} + 2 \times 0 = 2v_1 + 2v_2$$

or: $$v_1 + v_2 = \frac{13}{4} \qquad (1)$$

Newton's law of restitution gives:

$$e = \frac{5}{8} = \frac{v_2 - v_1}{\frac{13}{4} - 0}$$

or: $$v_2 - v_1 = \left(\frac{5}{8}\right)\left(\frac{13}{4}\right) = \frac{65}{32} \qquad (2)$$

Eliminating v_1 and finding v_2 by adding equations (1) and (2):

$$2v_2 = \frac{65}{32} + \frac{13}{4} = \frac{65 + 8 \times 13}{32} = \frac{169}{32}$$

So: $$v_2 = \frac{169}{64} = 2\frac{41}{64}$$

The speed of C after the second impact is $2\frac{41}{64}\,\mathrm{m\,s^{-1}}$.

Example 12

Particles A and B have mass 0.75 kg and 0.6 kg respectively. They are approaching each other moving along the same straight line with speeds of $6\,\text{m s}^{-1}$ and $4\,\text{m s}^{-1}$ respectively. The coefficient of restitution between them is $\frac{4}{5}$. After the collision the direction of motion of B is reversed. Particle B then strikes a fixed vertical wall at right angles. If the coefficient of restitution between B and the wall is $\frac{2}{3}$, show that B will collide again with A and find the speeds of A and B after this second collision.

1st impact

Before: $\rightarrow 6\,\text{m s}^{-1}$ A (0.75 kg) $\quad 4\,\text{m s}^{-1} \leftarrow$ B (0.6 kg) $\quad (e = \frac{4}{5})$

After: A $\rightarrow v_1\,\text{m s}^{-1}$ $\quad$ B $\rightarrow v_2\,\text{m s}^{-1}$

Using conservation of momentum gives:

$$(0.75) \times 6 - (0.6) \times 4 = (0.75)v_1 + (0.6)v_2$$

or:

$$5v_1 + 4v_2 = 14 \qquad (1)$$

Newton's Law of Restitution gives:

$$e = \frac{4}{5} = \frac{v_2 - v_1}{6 + 4}$$

or:

$$v_2 - v_1 = 10 \times \frac{4}{5} = 8 \qquad (2)$$

Solving equations (1) and (2) simultaneously gives:

$$v_1 = -2 \quad \text{and} \quad v_2 = 6$$

So after the 1st impact:

A: $2\,\text{m s}^{-1} \leftarrow$ $\quad$ B: $\rightarrow 6\,\text{m s}^{-1}$

taking account of the minus sign in v_1.

2nd impact

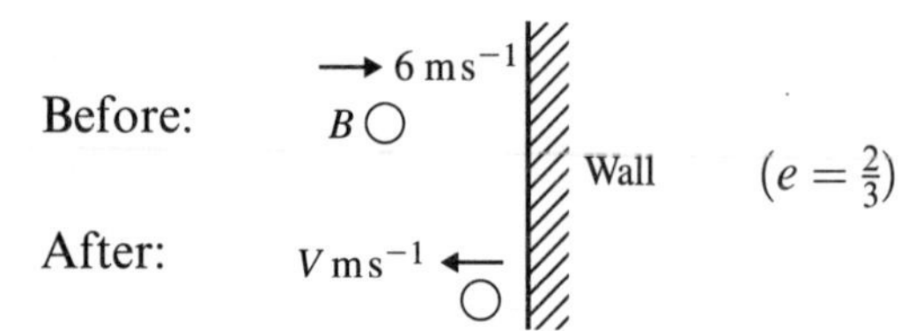

$(e = \frac{2}{3})$

By Newton's law of restitution:

$$e = \frac{V}{v} \quad \text{or} \quad V = ev$$

So: $$V = \frac{2}{3} \times 6 = 4$$

The rebound speed of B is $4\,\text{m}\,\text{s}^{-1}$.
2nd collision between A and B

$2\,\text{ms}^{-1} \leftarrow A \qquad 4\,\text{ms}^{-1} \leftarrow B \qquad (e = \frac{4}{5})$

As B is moving faster than A there will be a further collision between A and B.

Before: $2\,\text{ms}^{-1} \leftarrow$ A (0.75 kg) $\qquad 4\,\text{ms}^{-1} \leftarrow$ B (0.6 kg)

After: $\rightarrow V_1\,\text{ms}^{-1} \qquad \rightarrow V_2\,\text{ms}^{-1}$

Using conservation of momentum, taking the direction of V_1 and V_2 in the diagram as positive, gives:

$$-(0.75) \times 2 - (0.6)4 = (0.75)V_1 + (0.6)V_2$$

or: $$-26 = 5V_1 + 4V_2 \qquad (3)$$

Newton's law of restitution gives:

$$e = \frac{4}{5} = \frac{V_2 - V_1}{4 - 2}$$

or: $$V_2 - V_1 = 2 \times \frac{4}{5} = \frac{8}{5} \qquad (4)$$

Eliminating V_1 between equations (3) and (4) gives:

$$-18 = 9V_2$$

So: $$V_2 = -2$$

Substituting this value of V_2 into equation (4) gives:

$$V_1 = -3.6$$

So after this collision:

$3.6\,\text{ms}^{-1} \leftarrow A \qquad 2\,\text{ms}^{-1} \leftarrow B$

The speeds of A and B after the second collision are $3.6\,\text{m}\,\text{s}^{-1}$ and $2\,\text{m}\,\text{s}^{-1}$ respectively.

Exercise 4D

1 Each part of this question involves 3 small smooth spheres A, B and C of equal radius moving along the same straight line on a smooth horizontal plane. The sphere A collides with B, and then B collides with C. The diagrams show the situations

(i) before any collision,

(ii) after A and B have collided,

(iii) after B has collided with C.

In (a) $e = \frac{1}{2}$ between any two spheres in collision, and in (b) $e = \frac{3}{4}$ between any two spheres in collision.

In each part find x, y, v and w.

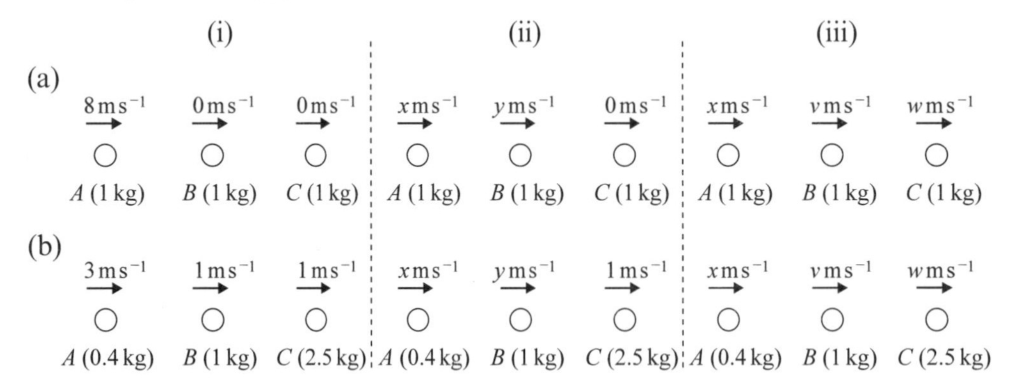

2 A small smooth sphere A of mass 2 kg moving on a smooth horizontal floor with speed $5\,\text{m s}^{-1}$ collides directly with a small smooth sphere of mass 1 kg which is at rest. The coefficient of restitution between the spheres is $\frac{4}{5}$.

(a) Find the speed of B after the impact.

Following the impact sphere B goes on to strike a vertical wall which is at right angles to the direction of motion of the spheres. The coefficient of restitution between sphere B and the wall is e. The sphere A is brought to rest by a second collision with B.

(b) Show that $e = \frac{2}{9}$.

3 Three small smooth spheres A, B and C have mass 0.1 kg, 0.2 kg and 0.4 kg respectively. They are at rest on a smooth table with their centres collinear and B between A and C. Sphere A is projected directly towards B with speed $8\,\text{m s}^{-1}$. After C has been struck by B it moves with a speed of $2\,\text{m s}^{-1}$. The coefficient of restitution between any two spheres is e.

(a) Show that $e = \frac{1}{2}$.

(b) Show that A and B are brought to rest by the collision.

4 Three perfectly elastic particles A, B and C with masses 3 kg, 2 kg and 1 kg respectively lie at rest in a straight line on a smooth horizontal table with B between A and C. Particle A is projected directly towards B with speed $5\,\text{m s}^{-1}$ and after A has collided with B, B collides with C.
(a) Find the speed of each particle after the second impact.
(b) State, giving a reason, whether there will be a third collision.

5 Three small smooth identical spheres each of mass 1 kg move on a straight line on a smooth horizontal floor. Initially B lies between A and C. Spheres A and B are projected directly towards each other with speeds $3\,\text{m s}^{-1}$ and $2\,\text{m s}^{-1}$ respectively and sphere C is projected directly away from sphere A with speed $2\,\text{m s}^{-1}$. The coefficient of restitution between any two spheres is e. Show that sphere B will only collide with sphere C if $e > \frac{3}{5}$. Describe a 'realistic situation' this could be used to model.

6 Three small smooth spheres A, B and C, of equal size and of mass $2m$, $7m$ and $14m$ respectively, are at rest on a smooth horizontal floor with their centres in that order on a straight line. The coefficient of restitution between each pair of spheres is $\frac{1}{2}$. Sphere A is projected so that it collides with sphere B.
(a) Show that, after two collisions have taken place, sphere B is at rest and spheres A and C are moving in opposite directions with equal speeds.
(b) Find what fraction of the original kinetic energy remains after the two collisions.

7 Two small smooth balls A and B of equal radius and mass 4 kg and 5 kg respectively lie at rest on a smooth horizontal floor. Ball A is projected with speed u and collides with ball B. Following this collision ball B then strikes a smooth vertical wall normally. After rebounding from the wall, ball B again collides with ball A. Given that ball B is brought to rest by this second collision with A,
(a) show that $2e^3 - 3e^2 - 3e + 2 = 0$,
where e is the coefficient of restitution between the two balls and between ball B and the wall.
(b) verify that $e = \frac{1}{2}$ is the only practical solution of the equation.

4.6 A collection of problems involving collisions

This section considers several problems where collisions occur at some point in the solution, but other mechanical principles discussed elsewhere in Books M1 and M2 are also involved.

Example 13

Two small smooth beads A and B of mass 0.5 kg and 0.6 kg respectively are threaded onto a smooth straight wire, fixed in a horizontal position, on which each bead is free to move. Initially the beads are 2 m apart. At $t = 0$ A is given a speed of $2.4\,\text{m s}^{-1}$ towards B and B is given a speed of $1.6\,\text{m s}^{-1}$ towards A.

In the subsequent motion A and B collide and A is brought to rest by the collision. Calculate:

(a) the time for which A is in motion
(b) the distance A moves
(c) the speed of B after the collision
(d) the kinetic energy lost in the collision
(e) the magnitude of the impulse exerted by B on A in the collision.

The initial situation is summarised in the following diagram.

→ $2.4\,\text{ms}^{-1}$ $1.6\,\text{ms}^{-1}$ ←

A (0.5 kg) B (0.6 kg)

(a) The closing speed of the beads is the speed of B as seen by A (or the speed of A as observed by B).

That is: $(2.4 + 1.6) = 4\,\text{m s}^{-1}$

The time to the collision is then $\frac{2}{4} = \frac{1}{2}$ s.

(b) This is the time that A is in motion, as it is brought to rest by the collision. In this time A moves a distance of

$$\tfrac{1}{2} \times (2.4) = 1.2\,\text{m}$$

(c) Taking the direction from left to right as the positive direction, the total linear momentum before the collision is:

$$(0.5 \times 2.4) - (0.6 \times 1.6)$$

Suppose after the collision B moves in the positive direction with speed v. Then the linear momentum after impact is $0.6\,v$.

The principle of conservation of momentum gives:

$$0.5 \times 2.4 - 0.6 \times 1.6 = 0.6v$$
$$0.5 \times 4 - 1.6 = v$$
$$v = 0.4$$

The speed of B after the collision is $0.4\,\text{m s}^{-1}$.

(d) The total K.E. before the collision is:

$$\tfrac{1}{2}(0.5)(2.4)^2 + \tfrac{1}{2}(0.6)(1.6)^2 = 1.44 + 0.768 = 2.208\,\text{J}$$

The total K.E. after the collision is:

$$0 + \tfrac{1}{2}(0.6)(0.4)^2 = 0.048\,\text{J}$$

The loss in kinetic energy is then 2.16 J.

(e) The magnitude of the impulse exerted by B on A is just the change in the momentum of A due to the collision. This is

$$0.5 \times 2.4 - 0 = 1.2\,\text{N s}$$

Example 14

Two particles A and B of mass $5m$ and $4m$ respectively are connected by a light inextensible string. The particles are placed side by side on a smooth horizontal floor and then A is projected, with speed V, directly away from B. When the string becomes taut particle B is jerked into motion and A and B then move with a common speed in the direction of the initial motion of A. Find:

(a) the common speed of the particles after the string becomes taut
(b) the magnitude of the impulse exerted on B when the string becomes taut
(c) the loss of total kinetic energy due to the jerk.

Before the jerk the situation is:

$0 \rightarrow$ $\quad$ $V \rightarrow$

$B\ (4m)$ $\quad$ $A\ (5m)$

Suppose the common speed of the particles after the jerk is v. Then after the jerk the situation is:

$v \rightarrow$ $\quad$ $v \rightarrow$

$B\ (4m)$ ———— $A\ (5m)$

(a) As explained in Book M1, the linear momentum is conserved during the jerk.

So: momentum before = momentum after

$$5mV = (5m + 4m)v$$

Hence:

$$v = \tfrac{5}{9}V$$

The common speed of the particles is $\frac{5}{9}V$.

(b) The impulse exerted on B is equal to the change in momentum of B.

So:

$$I = 4mv - 0 = 4mv$$
$$= 4m(\tfrac{5}{9}V) = 2\tfrac{2}{9}mV$$

The impulse is $2\frac{2}{9}mV \rightarrow$

(c) The K.E. before the jerk is $\frac{1}{2}(5m)V^2$

The K.E. after the jerk is $\frac{1}{2}(9m)v^2$

$$= \tfrac{9}{2}m\left(\tfrac{5}{9}V\right)^2$$
$$= \tfrac{25}{18}mV^2$$

Hence the loss in total kinetic energy is

$$mV^2\left(\tfrac{5}{2} - \tfrac{25}{18}\right) = \tfrac{20}{18}mV^2$$
$$= 1\tfrac{1}{9}mV^2$$

Example 15

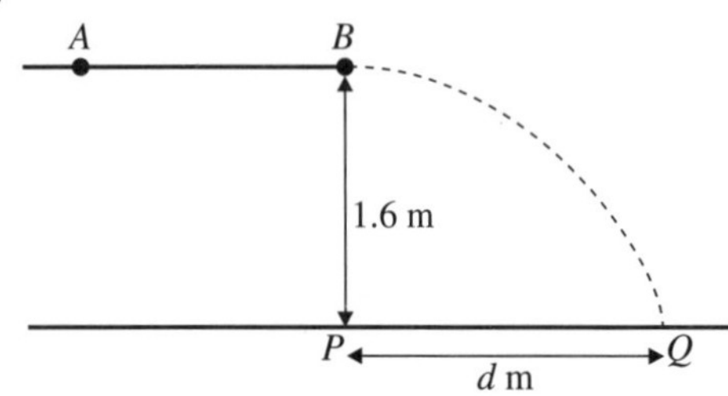

The diagram shows two particles A and B of mass 0.3 kg and 0.1 kg, respectively, initially at rest on a smooth horizontal shelf with B at the end of the shelf. The shelf is 1.6 m above a horizontal floor. At $t = 0$ particle A is projected towards B with speed $2.8\,\text{m s}^{-1}$. Particle A strikes particle B and they coalesce into a single particle C. The particle C then falls from the shelf and strikes the floor at a horizontal distance of d m from the point of leaving the shelf. Find d.

Using the principle of conservation of linear momentum:

$$\text{momentum before collision} = \text{momentum after collision}$$
$$0.3 \times 2.8 = (0.3 + 0.1)u$$

where u is the speed of particle C after the collision.

So: $$u = \frac{0.3 \times 2.8}{0.4} = 2.1\,\text{m s}^{-1}$$

When C leaves the shelf it moves as a projectile.

The time T taken by C to reach the floor is obtained from the equation $s = ut + \frac{1}{2}at^2$ for the vertical motion.

So: $$1.6 = 0 + \tfrac{1}{2}gT^2$$
$$T^2 = \frac{2 \times 1.6}{9.8} = \tfrac{16}{49}$$
$$T = \tfrac{4}{7}\,\text{s}$$

In this time the particle travels the horizontal distance PQ. This distance is just

$$\tfrac{4}{7} \times 2.1 = 1.2\,\text{m}$$

Example 16

Two particles A and B each of mass 2 kg are at rest 2 m apart on a rough horizontal floor. The coefficient of friction between the particles and the floor is $\frac{5}{12}$. Particle A is projected towards B with a speed of $7\,\text{m s}^{-1}$.

(a) Show that the speed of A just before it strikes B is $\sqrt{(\frac{98}{3})}\,\text{m s}^{-1}$.
Particle A collides with particle B and the two particles coalesce into a single particle C.
(b) Show that particle C travels a further 1 m before coming to rest.

(a) The forces acting on particle A while it is sliding are shown in the diagram.

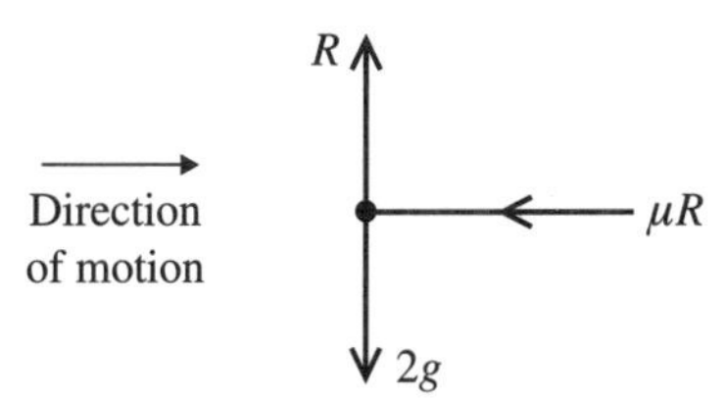

Resolving ↑ gives $R = 2g$.

So: $$\mu R = \tfrac{5}{12}(2g) = \tfrac{5}{6}g$$

Using $F = ma$ gives: $$2f = -\tfrac{5}{6}g$$

where f is the acceleration.

So: $$f = -\tfrac{5}{12}g \quad \text{(This is just } -\mu g.\text{)}$$

Using the constant acceleration equation, $v^2 = u^2 + 2as$, with $u = 7\,\text{m s}^{-1}$, $a = -\frac{5}{12}g$, $s = 2$, gives

$$v^2 = (7)^2 - 2(\tfrac{5}{12})g.2$$
$$= \tfrac{98}{3}$$

So the speed of A just before it strikes B is $\sqrt{(\frac{98}{3})}\,\text{m s}^{-1}$.

(b) When the two particles collide linear momentum is conserved, so:

$$2\sqrt{(\tfrac{98}{3})} + 0 = 4V$$

where V is the speed of the combined particle immediately after the collision.

So: $$V = \tfrac{1}{2}\sqrt{(\tfrac{98}{3})} = \sqrt{(\tfrac{49}{6})}\,\text{m s}^{-1}$$

The particle C will have the same deceleration of $\frac{5}{12}g$, as may be seen from the above analysis for A.

Using the constant acceleration formula $v^2 = u^2 + 2as$, with $v = 0$, $u = V$ and $a = -\frac{5}{12}g$, gives:

$$0 = \tfrac{49}{6} - 2(\tfrac{5}{12})gd$$

So: $$d = \tfrac{49}{6} \times \tfrac{6}{5} \times \tfrac{1}{9.8} = 1\,\text{m}$$

Example 17

A soldier fires a bullet horizontally from a rifle. The rifle has mass 4 kg and the bullet has mass 20 g. The initial speed of the bullet is 350 m s^{-1}.

(a) Find, in m s^{-1}, the initial speed with which the rifle recoils.
(b) Find, in J, the total kinetic energy generated as a result of firing the bullet.

The rifle is brought to rest by a horizontal force exerted by the soldier's shoulder, against which the rifle is pressed. This force is assumed to be constant. Given that the rifle recoils a distance of 2 cm before coming to rest,
(c) find the magnitude of the horizontal force exerted on the rifle by the soldier in bringing it to rest.

(a) For consistency of units both masses must be in kg. So the mass of the bullet is 0.02 kg.

Suppose v is the initial speed with which the rifle recoils.

Diagrammatically the situation is then:

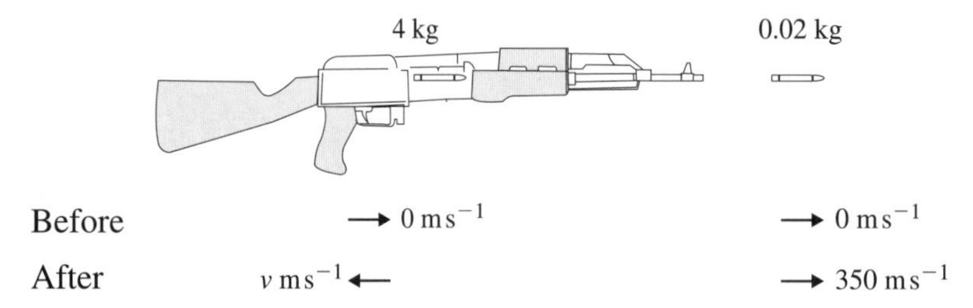

Before	$\rightarrow 0\text{ m s}^{-1}$	$\rightarrow 0\text{ m s}^{-1}$
After	$v\text{ m s}^{-1} \leftarrow$	$\rightarrow 350\text{ m s}^{-1}$

Conservation of linear momentum gives

$$4v = 0.02 \times 350$$

$$v = \frac{0.02 \times 350}{4} = 1.75$$

So the initial speed of the rifle is 1.75 m s^{-1}.

(b) The kinetic energy generated

$= (\text{K.E. of bullet}) + (\text{K.E. of rifle})$
$= \frac{1}{2}(0.02) \times (350)^2 + \frac{1}{2}(4) \times (1.75)^2$
$= 1225 + 6.125 = 1231.125\text{ J}$

The kinetic energy generated is 1230 J.

(c) Let the constant force have magnitude F N. Then the work done by the force in bringing the rifle to rest is $F \times (0.02)$ J, as $2\text{ cm} = 0.02\text{ m}$. This must be equal to the K.E. of the rifle, which is 6.125 J.

So:
$$F \times (0.02) = 6.125$$
$$F = 306.25$$

The magnitude of the force exerted by the solder is 306 N.

Exercise 4E

Whenever a numerical value of g is required take $g = 9.8\ \text{m s}^{-2}$.

1 A particle A of mass 2 kg is moving in a straight line on a horizontal table. It collides with another particle B of mass 1 kg moving in the same straight line on the table. Immediately before the collision the speed of A is $5\ \text{m s}^{-1}$ and the speed of B is $2\ \text{m s}^{-1}$ and the particles are moving in the same direction. In the collision the two particles coalesce to form a single particle C.

(a) Find the speed of C immediately after the collision.

The table is rough and the coefficient of friction between the table and C is $\frac{2}{7}$.

(b) Find the distance moved by C before coming to rest.

2 A ball A of mass 2 kg falls from rest and after falling for 1 s meets another ball B of mass 2 kg which is moving vertically upwards with a speed of $15\ \text{m s}^{-1}$. Given that the coefficient of restitution between the two balls is $\frac{3}{4}$,

(a) find the speed of A and the speed of B immediately after the collision, stating clearly their directions of motion,

(b) calculate the percentage loss in kinetic energy as a result of the collision.

3 A pile driver of mass 1200 kg falls from a height of 1.6 m on to a pile of mass 400 kg. After the impact the pile and the pile driver move on together. Given that the pile is driven a distance of 0.25 m into the ground, find:

(a) the speed at which the pile starts to move into the ground,

(b) the magnitude of the resistance of the ground, in kN (assumed constant).

4 A rocket of mass 75 kg is moving horizontally with speed $300\ \text{m s}^{-1}$ when an internal explosion causes it to split into two pieces, the head and the tail. Immediately after the explosion the head, which has mass 55 kg, moves with speed $450\ \text{m s}^{-1}$ in the original direction of motion. The tail, of mass 20 kg, also moves horizontally after the collision.

(a) Find the speed of the tail immediately after the explosion and state its direction of motion.

(b) Find the kinetic energy created by the explosion, giving your answers in kJ to 3 significant figures.

5 Two particles A and B each of mass 0.5 kg are connected by a light inextensible string of length 2 m. Initially they are at the same horizontal level, side by side, at rest. Particle B is then released.

(a) Find the speed of B just as the string becomes taut.

If A is released at the instant the string becomes taut, calculate:

(b) the common speed with which A and B together begin to move

(c) the impulsive tension in the string.

6 Two particles, A of mass $5m$ and B of mass $3m$, are connected by a light inextensible string. The particles are placed side by side on a smooth horizontal bench and A is projected directly away from B with speed u along the bench. At the instant when the string becomes taut, B is jerked into motion and A and B then move with the same speed in the direction initially taken by A. Find:

(a) the common speed of A and B after the string becomes taut

(b) the impulse exerted on B when the string becomes taut

(c) the loss in total kinetic energy when the string becomes taut. [E]

7 Two beads, A and B, of mass 0.05 kg and 0.06 kg respectively, are threaded on to a smooth straight fixed horizontal wire on which each bead is free to move. Initially the beads are placed on the wire at a distance 3.8 m apart. Instantaneously, A is given a speed of $1.2\,\text{m}\,\text{s}^{-1}$ towards B, and B is given a speed of $0.7\,\text{m}\,\text{s}^{-1}$ towards A.

The beads subsequently collide and A is brought to rest by the collision. Calculate:

(a) the time for which A is in motion

(b) the distance A moves

(c) the speed of B after the collision

(d) the kinetic energy lost in the collision, and state the units in which your answer is measured

(e) the magnitude of the impulse exerted by B on A in the collision, and state the units in which your answer is measured. [E]

8 A particle A of mass 0.7 kg is moving with speed $3\,\text{m}\,\text{s}^{-1}$ on a horizontal smooth table of height 0.9 m above a horizontal floor. Another particle B, of mass M kg, is at rest on the edge of the table top. Particle A strikes particle B, and they coalesce into a single particle C. The particle C then falls from the table. From the point of leavng the table to the point of hitting the floor, the horizontal displacement of C is 0.6 m.

(a) Show that C takes $\frac{3}{7}$ s to fall to the floor.

(b) Find the value of M. [E]

SUMMARY OF KEY POINTS

1 **Conservation of linear momentum**

When two particles collide:

total momentum before the collision = total momentum after the collision

$$m_1u_1 + m_2u_2 = m_1v_1 + m_2v_2$$

Before collision: m_1 ○ $\rightarrow u_1$ m_2 ○ $\rightarrow u_2$

After collision: $\rightarrow v_1$ $\rightarrow v_2$

2 **Newton's law of restitution**

$$\frac{\text{speed of separation of particles}}{\text{speed of approach of particles}} = e$$

where e is the coefficient of restitution between the particles.

3 Impact of a particle normally with a fixed surface.

Before impact:

After impact:

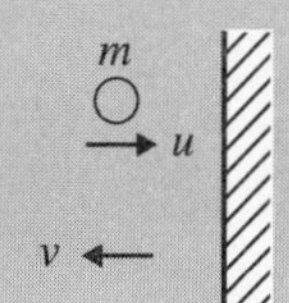

(speed of rebound) $= e$ (speed of approach)

where e is the coefficient of restitution between the particle and the surface.

Statics of rigid bodies

5

5.1 Moment of a force

The definition of the moment of a force about a point was given in chapter 6 of Book M1.

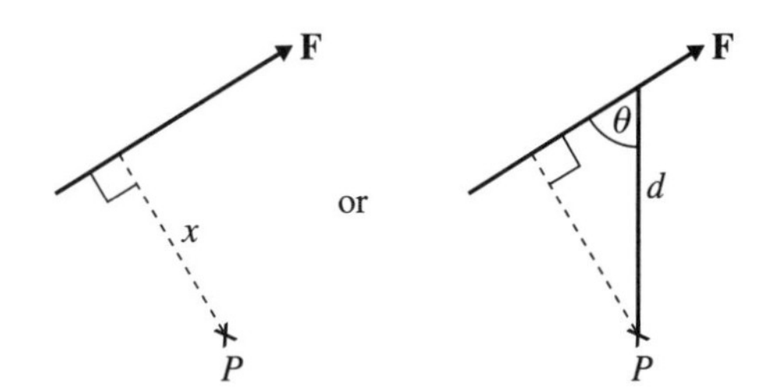

The moment of the force **F** about the point P is the product of the magnitude of the force **F** and the perpendicular distance from the point P to the line of action of **F**.

So moment of **F** about P in the above situation is:

$$Fx = Fd\sin\theta$$

The moment of a force is measured in newton metres (N m).

The moment of a force also has a sense of rotation, either clockwise ↷ or anticlockwise ↶. In the above case the moment of **F** about P is clockwise ↷.

The algebraic sum of moments

If a number of coplanar forces act on a rigid body then their moments about a given point may be added, provided that due regard is given to the *sense* of each moment.

Suppose a lamina is acted on by three coplanar forces, F_1, F_2 and F_3 as shown in the figure:

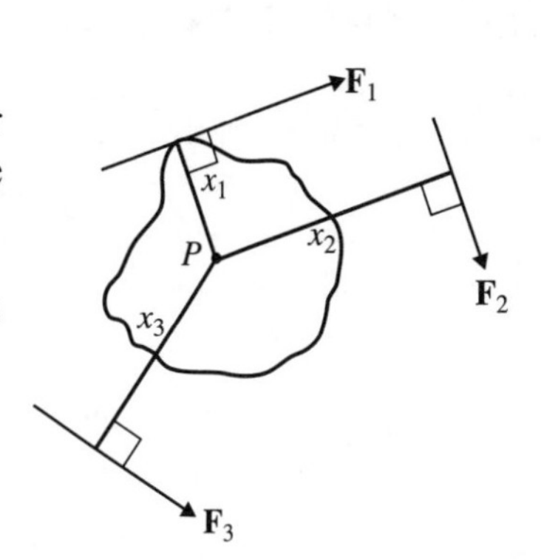

The total moment of the three forces is:

$$\curvearrowright \qquad F_1x_1 + F_2x_2 - F_3x_3$$

Example 1

A uniform rod AB of mass 5 kg and length 6 m is smoothly hinged at A and has a particle of mass 20 kg attached to it at B. A light inextensible string is attached to the rod at the point C where $AC = 4$ m and to the point D vertically above A, keeping the rod in a horizontal position. The tension in the string is T N. If the angle between the rod and the string is 30°, calculate in terms of T the resultant moment, about A, of the forces acting on the rod.

The information given is summarised in the following diagram:

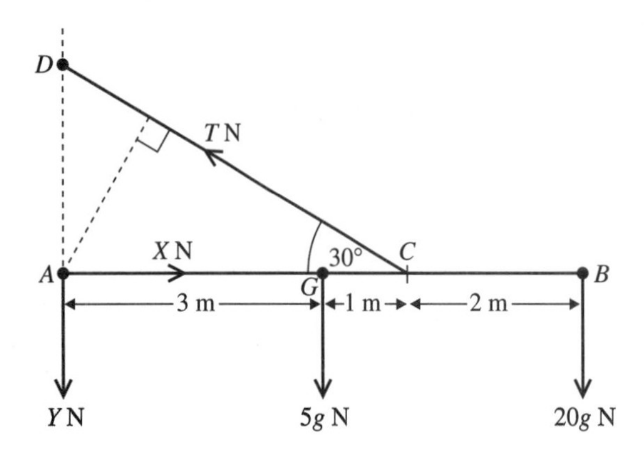

The forces acting on the rod are:
its weight acting at G
the weight of the particle acting at B
the tension in the string acting at C
the reaction of the hinge acting at A

Since the direction of the reaction at the hinge is not known we work with X N and Y N, the horizontal and vertical components.

The reaction of the hinge at A has no moment about A. The resultant moment of the other forces clockwise about A is:

$$\curvearrowright \quad [5g \times 3 + 20g \times 6 - T4\sin 30°]\,\text{N m} = [135g - 2T]\,\text{N m}$$

5.2 Equilibrium of rigid bodies

- **A rigid body is in equilibrium if**
  (i) the vector sum of the forces acting is zero; that is, the sum of the components of the forces in any given direction is zero
  (ii) the algebraic sum of the moments of the forces about any given point is zero.

In chapter 2 of this book you studied some simple cases of equilibrium in which all the forces acting on the body were parallel or anti-parallel. Now this condition is relaxed.

The following procedure is suggested for problems in which all the forces acting on the body are in the same plane. Forces which act in the same plane are called **coplanar forces**.

(1) **Draw** a clearly labelled diagram showing all the forces acting on the body.
(2) **Resolve** the forces in each of two perpendicular directions and equate the resolved parts to zero. (Almost always the best directions in which to resolve are horizontally and vertically or parallel and perpendicular to an inclined plane.)
(3) **Take moments** about any convenient point and equate the algebraic sum of the moments to zero. (Careful choice of the point used may greatly simplify the solution in a particular problem.)

Example 2

A uniform rod AB of length $2a$ and mass m is smoothly hinged at A. It is maintained in equilibrium by a horizontal force of magnitude P acting at B. Given that the rod is inclined at 30° to the horizontal with B below A find:

(a) the value of P
(b) the magnitude and direction of the reaction at the hinge.

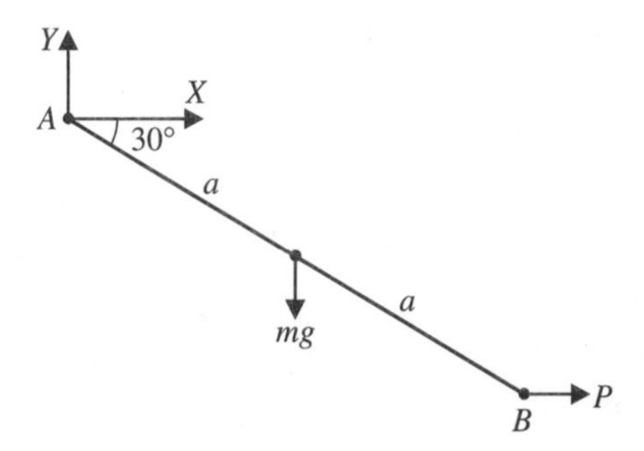

The reaction at the hinge is assumed to have horizontal and vertical components X and Y respectively.

Resolving horizontally: $\rightarrow X + P = 0$ (1)

Resolving vertically: $\uparrow Y - mg = 0$ (2)

As the forces X and Y act through A, taking moments about A will produce an equation not involving X and Y.

(a) Taking moments about $A2$

$$P \times 2a \sin 30° - mga \cos 30° = 0 \quad (3)$$

Using $\sin 30° = \frac{1}{2}$ and $\cos 30° = \frac{\sqrt{3}}{2}$ equation (3) gives:

$$Pa - \frac{\sqrt{3}}{2} mga = 0$$

So: $$P = \frac{\sqrt{3}}{2} mg \quad (4)$$

(b) From equation (2): $Y = mg$

Using equations (1) and (4):

$$X = -P = -\tfrac{\sqrt{3}}{2}mg$$

The reaction at the hinge is:

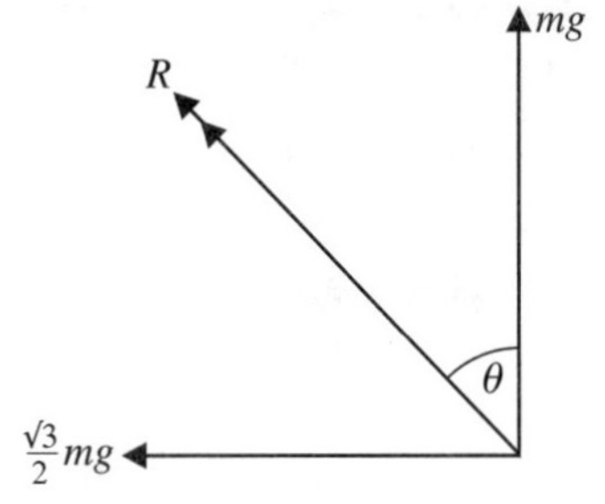

$$R = \sqrt{\left((mg)^2 + \left(\tfrac{\sqrt{3}}{2}mg\right)^2\right)}$$
$$= mg\sqrt{\left(1 + \tfrac{3}{4}\right)} = \tfrac{\sqrt{7}}{2}mg$$

And: $$\tan\theta = \frac{\tfrac{\sqrt{3}}{2}mg}{mg} = \tfrac{\sqrt{3}}{2}$$

so: $$\theta = 40.9^\circ$$

Hence the reaction at the hinge has magnitude $\tfrac{\sqrt{7}}{2}mg$ and makes an angle of 40.9° with the vertical.

Example 3

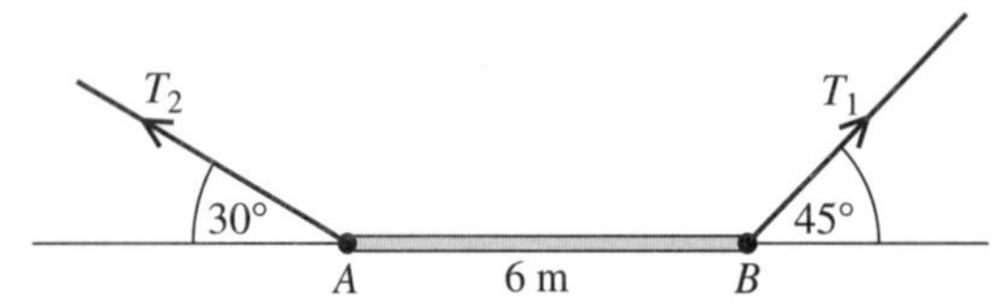

A non-uniform rod AB of mass 4 kg and length 6 m rests horizontally in equilibrium, supported by two strings attached at the ends A and B of the rod. The strings make angles of 30° and 45° with the horizontal as shown in the diagram.

(a) Obtain the tensions in each of the strings.
(b) Determine the position of the centre of mass of the rod.

Suppose the centre of mass G of the rod is a distance s m from A.

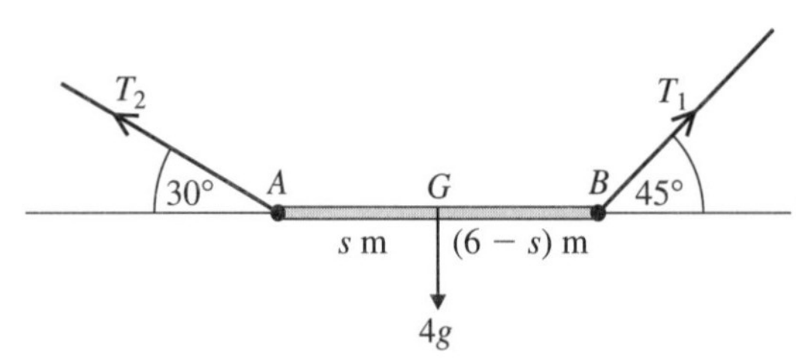

(a) Resolving horizontally: $\rightarrow$ $T_1\cos 45^\circ - T_2\cos 30^\circ = 0$ (1)
Resolving vertically: $\uparrow$ $T_2\sin 30^\circ + T_1\sin 45^\circ - 4g = 0$ (2)
Taking moments about A: $\curvearrowright$ $4gs - T_1 6\sin 45^\circ = 0$ (3)

Using $\cos 45^\circ = \tfrac{\sqrt{2}}{2}$ and $\cos 30^\circ = \tfrac{\sqrt{3}}{2}$ in equation (1) gives:

$$T_1\tfrac{\sqrt{2}}{2} - T_2\tfrac{\sqrt{3}}{2} = 0$$

so: $$T_2 = T_1\sqrt{\tfrac{2}{3}} \qquad (4)$$

Using $\sin 30° = \frac{1}{2}$ and $\sin 45° = \frac{\sqrt{2}}{2}$ in equation (2) gives:

$$T_2\left(\tfrac{1}{2}\right) + T_1\left(\tfrac{\sqrt{2}}{2}\right) - 4g = 0 \qquad (5)$$

Substituting for T_2 from (4) into (5) gives:

$$\tfrac{1}{2}\sqrt{\left(\tfrac{2}{3}\right)}T_1 + T_1\tfrac{\sqrt{2}}{2} = 4g$$

or:

$$\tfrac{1}{2}\sqrt{2}T_1\left[\tfrac{1}{\sqrt{3}} + 1\right] = 4g$$

So:

$$T_1 = \frac{8g}{\sqrt{2}\left(\frac{1}{\sqrt{3}} + 1\right)} = \frac{4g\sqrt{6}}{(\sqrt{3} + 1)} = 35.1\,\text{N} \qquad (6)$$

and from (4):

$$T_2 = \frac{8g}{\sqrt{3} + 1} = 28.7\,\text{N}$$

(b) From (3):

$$s = \frac{6T_1\sqrt{2}}{8g}$$

and using (6):

$$s = \frac{6\sqrt{3}}{\sqrt{3} + 1} = 3.80\,\text{m}$$

Exercise 5A

Whenever a numerical value of g is required take $g = 9.8\ \text{m s}^{-2}$.

1 A uniform beam AB of mass 20 kg and length 2 m is attached to a vertical wall by means of a smooth hinge at A. The beam is maintained in the horizontal position by means of a light inextensible string, one end of which is attached to the beam at B and the other end is attached to the wall at a point in the wall 2 m vertically above A.

(a) Calculate the tension in the string.

A particle of mass M kg is now attached to the beam at B. Given that the string is about to break, and that the breaking tension of string is 400 N,

(b) find the value of M to 3 significant figures.

2

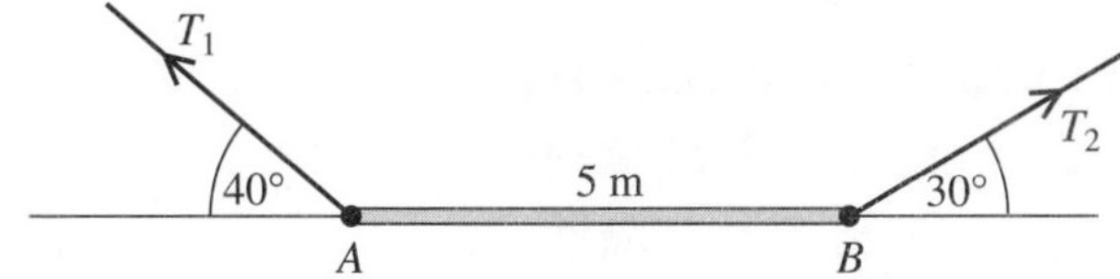

A non-uniform rod AB of mass 4 kg and length 5 m rests horizontally in equilibrium, supported by two strings attached at the ends A and B of the rod. The strings make angles of 40° and 30° with the horizontal as shown in the diagram.

(a) Obtain the tensions in each of the strings.

(b) Determine the position of the centre of mass of the rod.

3

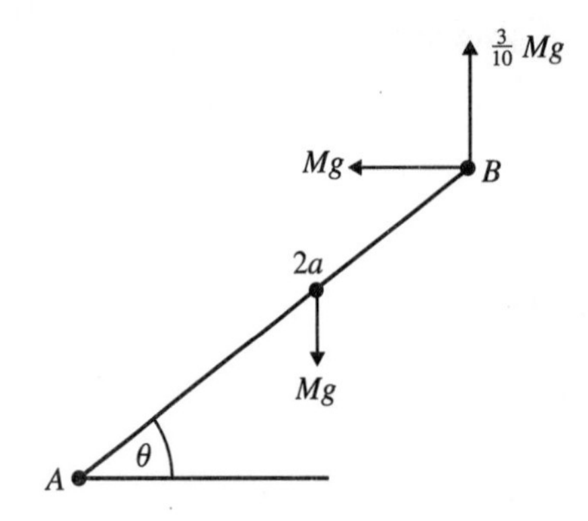

A uniform straight rod AB has mass M and length $2a$. The end A is smoothly hinged at a fixed point so that the rod can turn freely in a vertical plane. Horizontal and vertical forces of magnitudes Mg and $\frac{3}{10}Mg$ respectively are applied to the end B. These forces and the weight of the rod are shown in the diagram. The rod rests in equilibrium at an angle θ to the horizontal.

(a) By taking moments about A, find the value of $\tan\theta$.

(b) Calculate the magnitude, in terms of M and g, and the direction, to the nearest degree, of the force exerted by the hinge on the rod AB. [E]

4

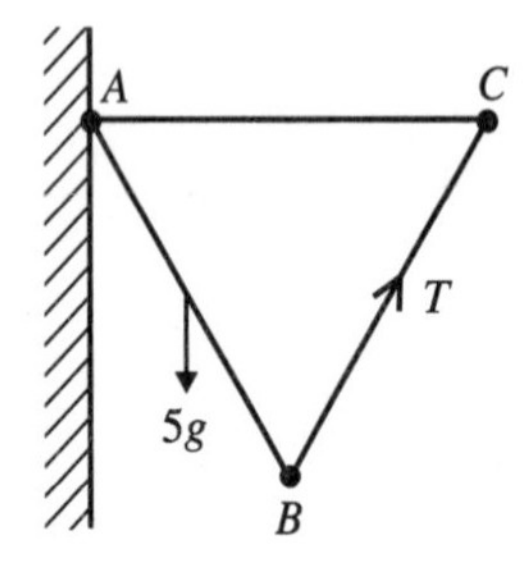

The end A of a uniform rod AB of length 2 m and mass 5 kg is freely hinged to a vertical wall and the other end B of the rod is supported by an inelastic string BC of length 2 m. The other end C of the string is attached to a point on the same horizontal level as A with $AC = 2$ m and such that A, B and C lie in the same vertical plane, as shown in the diagram.

(a) Find the tension T in the string.

(b) Determine the horizontal and vertical components of the reaction at the hinge.

5 $ABCD$ is a uniform square lamina of side 4 m and mass 8 kg. It is hinged at A so that it is free to move in a vertical plane. It is maintained in equilibrium with B vertically below A by a horizontal force acting at C and a vertical force acting at D, each of magnitude F newtons. Find:

(a) the value of F

(b) the magnitude and direction of the force exerted by the hinge on the lamina.

6 A uniform rod AB of length 4 m and mass 4 kg is freely hinged at A to a vertical wall. A force P is applied at B at right angles to the rod in order to keep the rod in equilibrium at an angle of 30° to the horizontal, as shown in the diagram.

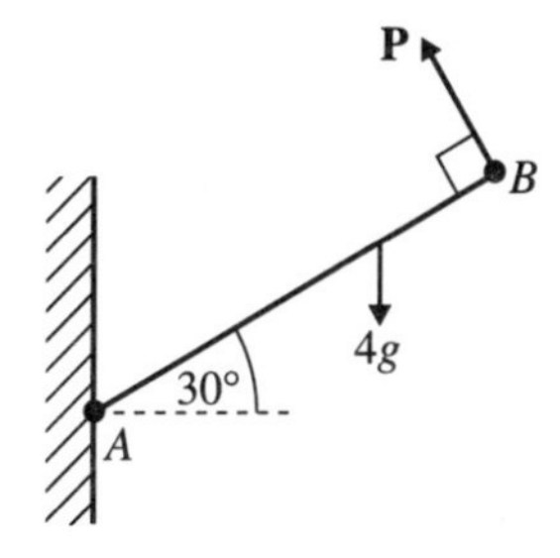

Find:

(a) the magnitude of **P**

(b) the magnitude of **R**, the reaction at the hinge.

7

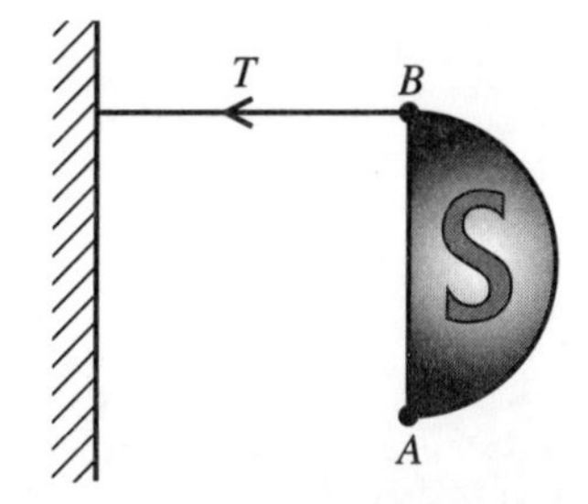

The diagram shows a new sign placed outside a shop. The sign is a semicircular metal plate of radius 1 m which is freely hinged at A so that it can move in a vertical plane. The plate is of mass 5 kg and is maintained in equilibrium, with the boundary diameter AB vertical and B above A, by a horizontal cable attached at B and in the vertical plane of the plate. By modelling the plate by a uniform semicircular lamina, find:

(a) the tension T in the cable

(b) the horizontal and vertical components of the force exerted by the hinge on the plate.

5.3 Limiting equilibrium

When one of the forces acting on a body is a frictional force and the body is in equilibrium it is necessary to be clear whether the body is in **limiting equilibrium**. (Limiting equilibrium is discussed in chapter 4 of Book M1.) It is to be stressed that **only in the case of limiting equilibrium**, when motion is on the point of taking place, **does the frictional force F have its maximum value μR** (where μ is the coefficient of friction).

Example 4

A uniform rod AB, of mass 10 kg, rests with its lower end A on a rough horizontal floor. The coefficient of friction between the rod and the floor is μ. One end of a string is attached to end B of the rod and the other end is fixed so that the string is perpendicular to the rod. The rod rests in equilibrium at an angle of 30° to the horizontal. Find:

(a) the tension T in the string

(b) the magnitudes of R the normal contact force and F the frictional force.

(c) Determine the least possible value of μ for equilibrium to be possible.

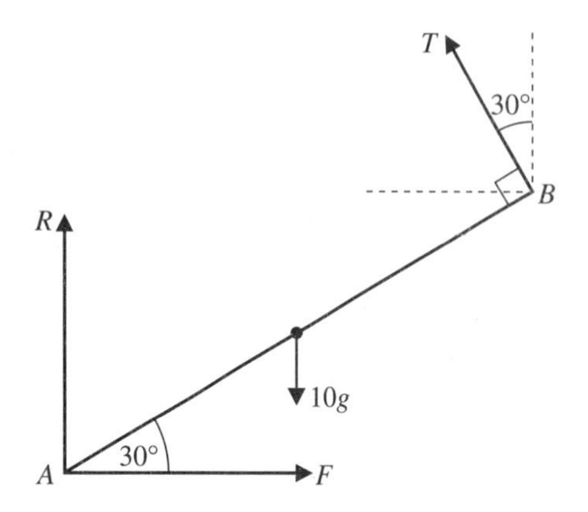

The diagram includes all the given information together with all the forces acting. Since A would move to the left if slipping occurred, the frictional force must act towards the right.

Resolving horizontally: $\rightarrow F - T\sin 30° = 0$ (1)

Resolving vertically: $\uparrow R + T\cos 30° - 10g = 0$ (2)

Let the rod have length $2a$.

Since both R and F act through the point A the simplest possible moment equation is obtained by taking moments about A.

Taking moments about A: $\quad 2\,T2a - 10ga\cos 30^\circ = 0 \quad (3)$

From (3): $\quad 2T - 10g\frac{\sqrt{3}}{2} = 0$ using $\cos 30^\circ = \frac{\sqrt{3}}{2}$

(a) So: $\quad T = \dfrac{5\sqrt{3}g}{2} = 42.4\,\text{N} \quad (4)$

(b) Using this result in (1):

$$F = T\sin 30^\circ = T(\tfrac{1}{2}) \text{ using } \sin 30^\circ = \tfrac{1}{2}$$
$$= 21.2\,\text{N}$$

Using (4) in (2):

$$R = 10g - T\cos 30^\circ$$
$$= 10g - \frac{T\sqrt{3}}{2} = 10g - \frac{5 \times 3}{4}g$$
$$= 61.3\,\text{N}$$

(c) Since $\mu = \dfrac{F_{\text{max}}}{R}$ the least value of μ necessary is $\dfrac{21.2}{61.3} = 0.346$ and the rod would then be in limiting equilibrium.

Problems involving ladders

The situation of a ladder resting against a wall with the foot of the ladder on the ground occurs frequently.

It may be modelled by a uniform rod in contact with a vertical wall and resting on horizontal ground. There are some circumstances where the frictional forces on the rod due to the wall or the floor are so small that they may, as a first model, be ignored; that is the contacts may be regarded as 'smooth'. In other cases the frictional forces cannot be neglected and so the contacts will be 'rough'.

When the rod is in contact with a **smooth** surface there will only be a normal contact force R, so $F = 0$. When the rod is in contact with a **rough** surface the frictional force F acts parallel to the surfaces in contact in such a direction as to oppose subsequent motion.

Example 5

A uniform ladder of mass 10 kg and length 6 m rests against a smooth vertical wall with its lower end on rough ground. The ladder rests in equilibrium at an angle of 60° to the horizontal. Find:

(a) the magnitude of the normal contact force S at the wall,

(b) the magnitude of the normal contact force R at the ground and the frictional force at the ground.

(c) Obtain the least possible value of μ.

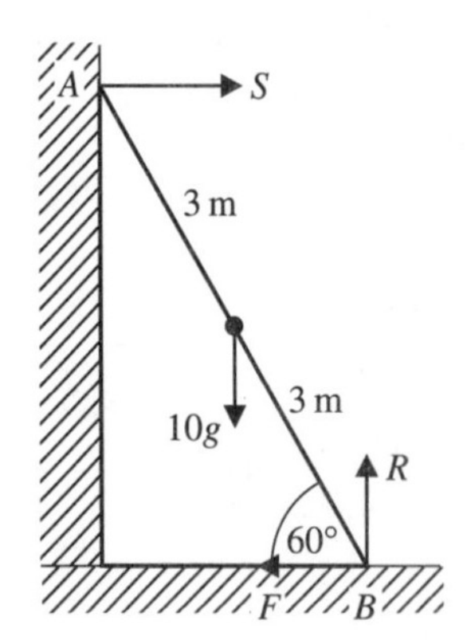

The ladder is modelled by a uniform rod. The contact at A is perfectly smooth and the contact at B is rough.

Resolving horizontally: $\rightarrow S - F = 0$ (1)

Resolving vertically: $\uparrow R - 10g = 0$ (2)

Moments about B: $\circlearrowleft S \times (6 \sin 60°) - 10g \times (3 \cos 60°) = 0$ (3)

Using $\sin 60° = \frac{\sqrt{3}}{2}$ and $\cos 60° = \frac{1}{2}$:

$$S3\sqrt{3} - 10g \times \tfrac{3}{2} = 0$$

(a) So: $S = \dfrac{10g}{2\sqrt{3}} = 28.3\,\text{N}$

(b) From (1): $F = S = 28.3\,\text{N}$

From (2): $R = 10g = 98\,\text{N}$

(c) Since: $\mu = \dfrac{F_{\text{max}}}{R}$

and: $\dfrac{F}{R} = \dfrac{28.3}{98} = 0.289$

μ must be at least 0.289.

Climbing a ladder

It is very important to know whether or not it is safe to climb to the top of a ladder. The answer will depend on the magnitude of the frictional force acting on the ladder. This in turn depends on the roughness of the ground on which the ladder rests. If the ladder is found to be in limiting equilibrium when a person is only part way up the ladder then any further ascent will cause the ladder to slip. Such a problem is dealt with in the next example, where the ladder is modelled by a uniform rod and the person by a particle at the point where he stands on the ladder.

Example 6

A uniform ladder of mass 30 kg and length 10 m rests against a smooth vertical wall with its lower end on rough ground. The coefficient of friction between the ground and the ladder is 0.3. The ladder is inclined at an angle θ to the horizontal where $\tan\theta = 2$. Find how far a boy of mass 30 kg can ascend the ladder without it slipping.

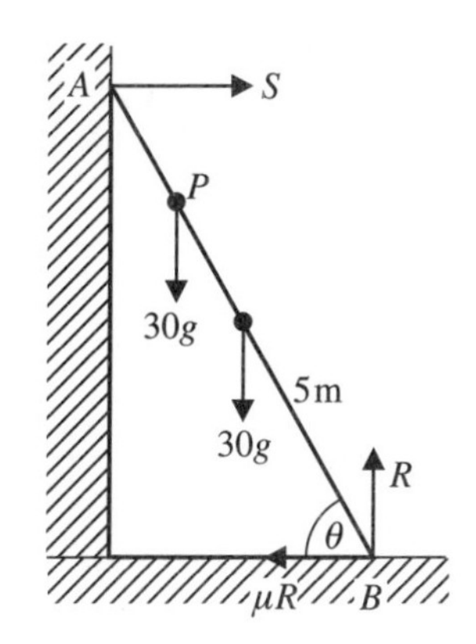

Let the boy be at the point P, where $BP = l$ m, when the ladder is about to slip.

Since the ladder is about to slip the frictional force is μR.

Resolving horizontally: $\rightarrow$ $S - \mu R = 0$ (1)

Resolving vertically: $\uparrow R - 30g - 30g = 0$ (2)

Taking moments about B:

$$S \times 10\sin\theta - 30g \times (5\cos\theta) - 30g \times (l\cos\theta) = 0 \quad (3)$$

From equation (2): $R = 60g$

Substituting into (1) with $\mu = 0.3$:

$$S = (0.3)60g = 18g \quad (4)$$

Using (4) in equation (3) gives:

$$18 \times 10g\sin\theta - 30 \times 5g\cos\theta - 30g \times l\cos\theta = 0$$

So: $6\tan\theta - 5 = l$

Using $\tan\theta = 2$ gives $l = 7$.

So the boy can ascend 7 m up the ladder before it begins to slip.

Example 7

A uniform rod AB of mass 20 kg and length 8 m rests with the end A on rough horizontal ground. The rod rests against a smooth peg C, where $AC = (4 + x)$ m. The rod is in limiting equilibrium inclined at an angle of 45°. Given that the coefficient of friction between the rod and the ground is $\frac{1}{2}$, find:

(a) the value of x
(b) the magnitude of the reaction at C.

The force acting at C is perpendicular to the rod since the peg is smooth. Since the rod is in limiting equilibrium the frictional force at the lower end of the rod is μR, where R is the normal contact force acting there.

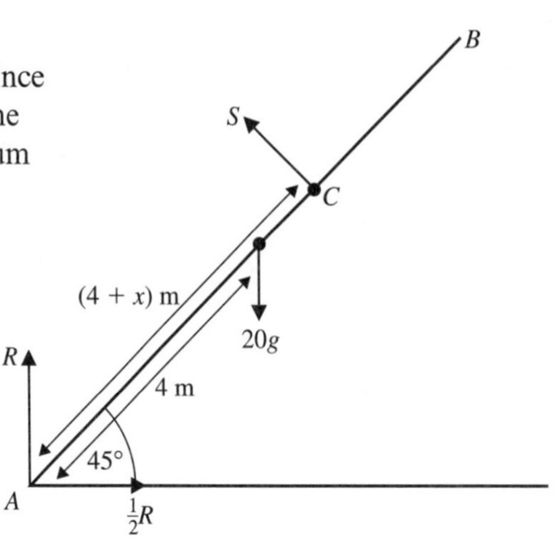

Resolving horizontally: $\rightarrow \quad \frac{1}{2}R - S\sin 45^\circ = 0$ (1)

Resolving vertically: $\uparrow R + S\cos 45^\circ - 20g = 0$ (2)

Taking moments about A: $2\ S(4 + x) - 20g \times 4\cos 45^\circ = 0$ (3)

From (1): $\frac{1}{2}R - S\frac{\sqrt{2}}{2} = 0$ since $\sin 45^\circ = \frac{\sqrt{2}}{2}$

So: $$S = \frac{R}{\sqrt{2}} \quad (4)$$

From (2): $R + S\frac{\sqrt{2}}{2} - 20g = 0$ since $\cos 45^\circ = \frac{\sqrt{2}}{2}$

Substituting for S from (4):

$$R + \frac{R}{2} - 20g = 0$$

So: $$\frac{3}{2}R = 20g$$

and: $$R = \frac{40g}{3}$$

and using (4): $$S = \frac{40g}{3\sqrt{2}} = 92.4\,\text{N} \quad (5)$$

From (3): $S(4+x) = 20g \times \frac{4\sqrt{2}}{2}$ since $\cos 45° = \frac{\sqrt{2}}{2}$

Using (5): $\frac{40g}{3\sqrt{2}}(4+x) = 20g \times \frac{4\sqrt{2}}{2}$

So: $4 + x = 6$

and: $x = 2$

(a) The value of x is 2.

(b) The reaction of C is of magnitude 92.4 N.

Exercise 5B

Whenever a numerical value of g is required take $g = 9.8\ \text{m s}^{-2}$.

1 A uniform ladder rests in limiting equilibrium with its top against a smooth vertical wall and its base on a rough horizontal floor. The coefficient of friction between the ladder and the floor is μ. Given that the ladder makes an angle θ with the floor show that:

$$2\mu \tan\theta = 1$$

2 A uniform ladder of mass 30 kg and length 10 m rests with one end on a smooth horizontal floor and the other end against a smooth vertical wall. The ladder is held in this position by a light inextensible string of length 5 m which has one end attached to the bottom of the ladder and the other end fastened to a point at the base of the wall vertically below the top of the ladder. Find the tension in the string.

3 A uniform ladder AB of mass 30 kg rests with its top A against a smooth vertical wall and its base B on rough horizontal ground. A mass of 30 kg is attached to the ladder at A. The coefficient of friction between the ladder and the ground is $\frac{3}{4}$. Given that the ladder is just about to slip, find the inclination of the ladder to the horizontal.

4 A uniform ladder of mass 30 kg and length 10 m rests with its top against a smooth vertical wall and its base on rough horizontal ground. The ladder rests in equilibrium at 60° to the horizontal with a man of mass 90 kg standing on the ladder at a point 7.5 m from its base.

(a) Find the magnitude of the normal contact force and the frictional force exerted on the ladder by the ground.

(b) Find the minimum value of the coefficient of friction between the ladder and the ground that would enable the man to climb to the top of the ladder.

5 A uniform ladder rests with one end on rough horizontal ground and the other end against a rough vertical wall. The coefficient of friction between the ground and the ladder is $\frac{3}{5}$ and the coefficient of friction between the wall and the ladder is $\frac{1}{3}$. The ladder is on the point of slipping when it makes an angle α with the horizontal. Find $\tan \alpha$.

6 A uniform rod AB of mass 40 kg and length 10 m rests with the end A on rough horizontal ground. The rod rests against a smooth peg C where $AC = 8$ m. The rod is in limiting equilibrium at an angle of $\alpha°$ to the horizontal. Given that the reaction at C is of magnitude 196 N find:

(a) the value of α

(b) the coefficient of friction between the rod and the ground.

7 A uniform ladder of mass 30 kg is placed with its base on rough horizontal ground. The coefficient of friction between the ladder and the ground is $\frac{1}{4}$. The upper end of the ladder rests against a smooth vertical wall, the ladder making an angle of 60° to the horizontal. Find the magnitude of the minimum horizontal force that must be applied to the base of the ladder to prevent slipping.

8 A uniform rod of mass 20 kg and length 20 m rests in limiting equilibrium with the upper end against a smooth vertical wall and the lower end on a rough horizontal floor.

(a) Given that the coefficient of friction is $\frac{1}{2}$, find the angle made by the ladder to the horizontal.

(b) Describe a 'physical situation' for which the above could be used as a model and state what assumptions would be made in using this model.

(c) How could this model be refined to take into account other attributes of the 'physical situation'?

9 A uniform ladder of mass M rests in limiting equilibrium with one end on rough horizontal ground and the other end against a rough vertical wall. The coefficient of friction between the ladder and the ground is μ and the coefficient of friction between the ladder and the wall is μ'. Given that the ladder makes an angle α with the horizontal, show that:

$$\tan\alpha = \frac{1 - \mu\mu'}{2\mu}$$

10 A uniform ladder AB, of mass 30 kg, rests in equilibrium with the end A in contact with rough horizontal ground and the end B against a smooth vertical wall. The vertical plane containing AB is at right angles to the wall and AB is inclined at 60° to the horizontal. The coefficient of friction between the ladder and the ground is μ.

(a) Find the magnitude of the force exerted by the wall on the ladder.

(b) Show that $\mu \geqslant \frac{1}{6}\sqrt{3}$.

A load of mass m is attached to the ladder at B.

(c) Given that $\mu = \frac{1}{5}\sqrt{3}$ and that the equilibrium is limiting, find m.

SUMMARY OF KEY POINTS

1 A rigid body is in **equilibrium** if:

(i) The vector sum of the forces acting is zero, that is the sum of the components of the forces in any given direction is zero.

(ii) the algebraic sum of the moments of the forces about any given point is zero.

2 Only in the case of limiting equilibrium, when motion is on the point of taking place, does the frictional force F have its maximum value μR.

Review exercise 2

Whenever a numerical value of g is required take $g = 9.8\,\text{m}\,\text{s}^{-2}$.

1 A truck P, of mass M, is moving with speed U on horizontal rails. It collides with a truck Q, of mass $4M$, which is at rest. The trucks move on together after the collision.

(a) Find the common speed of the trucks after their collision giving your answer in terms of U.

(b) Find the loss in kinetic energy due to the collision giving your answer in terms of M and U. [E]

2 Particles A and B have masses kM and M respectively, where k is a positive constant. The particles are moving in the same direction, along the same straight line, with speeds $2u$ and u respectively. There is a collision between A and B, which halves the speed of A. After the collision, both A and B continue to move in the same direction as before.

(a) Find, in terms of k and u, the speed of B after the collision.

In the collision, one-twelfth of the total kinetic energy is lost.

(b) Find the possible values of k. [E]

3 Two particles P and Q, each of mass m, are moving in the same direction along the same straight line with constant speeds $5u$ and u respectively. The particles collide and, after the collision, continue to move in the same direction as before, the speed of Q now being twice the speed of P.

(a) Show that the kinetic energy lost in the collision is $3mu^2$.

(b) Find the magnitude of the impulse exerted by P on Q in the collision. [E]

4 Particle A, of mass $3m$, moving with speed u on a horizontal plane, strikes directly a particle B, of mass m, which is at rest on the plane. The coefficient of restitution between A and B is e.

(a) Find, in terms of e and u, the speeds of A and B immediately after their collision.

Given that the magnitude of the impulse exerted by A on B is $\frac{5mu}{4}$,

(b) show that $e = \frac{2}{3}$.

Particle B then slides a distance $\frac{3u^2}{g}$ and strikes directly particle C, of mass λm, which is at rest on the plane. The coefficient of restitution between B and C is $\frac{2}{3}$ and the coefficient of friction between B and the plane is $\frac{1}{4}$. By considering the work done against friction, or otherwise,

(c) show that B strikes C with speed $\frac{1}{4}u$.

Given that B comes to rest immediately after striking C,

(d) find the value of λ. [E]

5 Two smooth spheres, A and B, of equal radii and masses m and $4m$ respectively, are moving on the surface of a smooth horizontal table. The sphere A, moving with speed u, overtakes and strikes directly the sphere B which is moving in the same direction with speed λu, where $0 < \lambda < 1$. The sphere A is brought to rest by the impact. Show that e, the coefficient of restitution between A and B, is given by

$$e = \frac{4\lambda + 1}{4(1 - \lambda)}.$$

Deduce that $\lambda \leqslant \frac{3}{8}$.

Given further that 25% of the total initial kinetic energy is lost in the collision between A and B, prove that

$$\lambda = (\sqrt{6} - 2)/2$$ [E]

6 A small smooth sphere S, of mass 0.2 kg, moving with speed $5\,\text{m}\,\text{s}^{-1}$ overtakes and collides with a second small smooth sphere T, of mass 0.3 kg, which is moving with speed $2\,\text{m}\,\text{s}^{-1}$ in the same direction as S. The coefficient of restitution between the spheres is 2/3. Calculate

(a) the speed of each sphere immediately after the impact,

(b) the magnitude of the impulse received by each sphere on impact.

7 A particle of mass m moving in a straight line with speed u receives an impulse of magnitude I in the direction of its motion. Show that the increase in kinetic energy is given by

$$I(I + 2mu)/(2m)$$ [E]

8 Three particles A, B and C have masses m, $3m$ and λm respectively. The particles lie at rest on a smooth horizontal plane in a straight line with B between A and C. Particle A is given a horizontal impulse, of magnitude J, and collides directly with B. After this collision A is at rest and B moves towards C with speed u. The coefficient of restitution at each impact is e.

(a) Find J in terms of m and u.

(b) Show that $e = \frac{1}{3}$.

(c) Find in terms of m and u, the loss in kinetic energy in the collision between A and B.

Particle B, moving with speed u, collides directly with particle C.

(d) Find, in terms of λ and u, the speeds of B and C after their collision.

(e) Show that A and B will have a second collision provided that $\lambda > 9$.

(f) Given that $\lambda = 6$, find, in terms of m and u, the magnitude of the impulse on B in the collision between B and C. [E]

9 The particles A and B of mass $3m$ and $2m$ respectively are moving due east in the same straight line on a smooth horizontal plane. They collide directly. The coefficient of restitution between the particles is $\frac{3}{7}$. After the collision, A and B continue to move due east in the same line with speed $6u$ and $9u$ respectively.

(a) Calculate the speeds of A and B before the collision.

(b) Calculate the magnitude of the impulse exerted by A on B due to the collision.

(c) Show that the kinetic energy lost in the collision is $24mu^2$.

[E]

10 A sphere of mass m moving along a smooth horizontal table with speed V collides directly with a stationary sphere of the same radius and of mass $2m$. Obtain expressions, in terms of V and the coefficient of restitution e, for the speeds of the two spheres after impact.

Half of the kinetic energy is lost in the impact. Find the value of e. [E]

11 The coefficient of restitution between two particles A and B is e, where $0 < e < 1$. The masses of A and B are m and em respectively. The particles are moving with constant speeds u and eu in the same horizontal line and in the same direction, as shown in the diagram, and they collide.

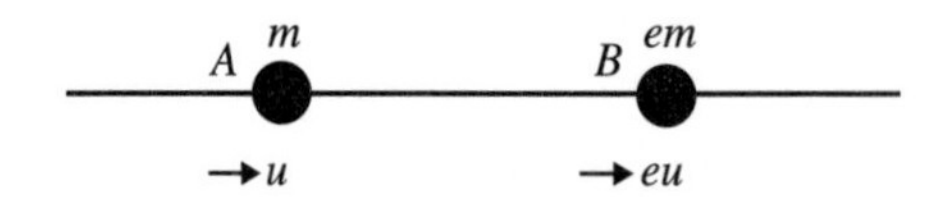

(a) Show that after the collision the speed of A is $u(1 - e + e^2)$ and that the speed of B is independent of e.
(b) Find the value of e for which the speed of A after the collision is least and deduce that, in this case, the total loss of kinetic energy due to the collision is $\frac{1}{32}mu^2$.
(c) Find the possible values of e for which the impulse of the force exerted by B on A due to the collision has magnitude $\frac{6}{25}mu$. [E]

12 Two spheres A and B, of equal radius and of mass m and $5m$ respectively, are moving on the surface of a smooth horizontal table. The sphere A, moving with speed $6u$, strikes directly the sphere B, which is initially moving in the same direction as A with speed u. After the impact, the speed of A is $2u$ and its direction of motion is reversed.
(a) Find the coefficient of restitution between A and B.
(b) Show that the kinetic energy lost in the impact is $1.6mu^2$. [E]

13 A particle P, of mass $2m$, is moving in a straight line with speed u at the instant when it collides directly with a particle Q, of mass m, which is at rest. The coefficient of restitution between P and Q is e.
(a) Show that, after the collision, P is moving with speed

$$\frac{(2-e)}{3}u$$

(b) Show that the loss of kinetic energy due to the collision is

$$\frac{mu^2}{3}(1-e^2)$$

(c) Find, in terms of m, u and e, the impulse exerted by P on Q in the collision. [E]

14 Two small smooth spheres, A and B, of equal radii and of masses m and $2m$ respectively, are moving on the surface of a smooth horizontal table. Initially, A and B are moving in the same direction with speeds $3u$ and $2u$ respectively. There is a direct collision of A and B, the coefficient of restitution between them being e.

(a) Find, in terms of u and e, the speeds of A and B after the collision.

(b) Show that, whatever the value of e, the speed of B after the collision cannot exceed $\frac{8}{3}u$. Given that $e = \frac{4}{5}$,

(c) find the magnitude of the impulse exerted by A on B in the collision. [E]

15 Two particles, A and B of mass m and $4m$ respectively, are placed on a smooth horizontal plane. Particle A is made to move on the plane with speed u so as to collide directly with B, which is at rest. After the collision B moves with speed ku, where k is a constant.

(a) Find the speed of A after the collision.

(b) By using Newton's law of restitution, show that

$$\frac{1}{5} \leqslant k \leqslant \frac{2}{5}$$ [E]

16 A sphere P, of mass m, is moving in a straight line with speed u on the surface of a smooth horizontal table. Another sphere Q, of mass $5m$ and having the same radius as P, is initially at rest on the table. The sphere P strikes the sphere Q directly, and the direction of motion of P is reversed by the impact. The coefficient of restitution between P and Q is e.

(a) Find an expression, in terms of u and e, for the speed of P after the impact.

(b) Find the set of possible values of e. [E]

17 A small ball B, of mass 120 g, moving with speed $14\,\text{m}\,\text{s}^{-1}$ collides directly with a small ball C, of mass 100 g, moving with speed $16\,\text{m}\,\text{s}^{-1}$ in the opposite direction. The coefficient of restitution between B and C is e. Immediately after the collision the speed of C is $v\,\text{m}\,\text{s}^{-1}$.

(a) Show that

$$v = \frac{1}{11}(180e + 4)$$

(b) Show that

$$\frac{4}{11} \leqslant v \leqslant 16\frac{8}{11}$$

(c) Find, in terms of e, the velocity of B.

(d) Given that B is brought to rest by the collision, find the value of e. [E]

18 A particle A of mass m, moving with speed u on a smooth horizontal surface, collides directly with a stationary particle B of mass $3m$.
The coefficient of restitution between A and B is e. The direction of motion of A is reversed by the collision.
(a) Show that the speed of B after the collision is $\frac{1}{4}u(1+e)$.
(b) Find the speed of A after the collision.
Subsequently, B hits a wall fixed at right angles to the direction of motion of A and B.
The coefficient of restitution between B and the wall is $\frac{1}{2}$.
After B rebounds from the wall, there is another collision between A and B.
(c) Show that $\frac{1}{3} < e < \frac{3}{5}$.
(d) In the case $e = \frac{1}{2}$, find the magnitude of the impulse exerted on B by the wall. [E]

19 A particle A of mass $2m$, moving with speed $2u$ in a straight line on a smooth horizontal table, collides with a particle B of mass $3m$, moving with speed u in the same direction as A.
The coefficient of restitution between A and B is e.
(a) Show that the speed of B after the collision is

$$\frac{1}{5}u(7+2e)$$

(b) Find the speed of A after the collision, in terms of u and e.
The speed of A after the collision is $\frac{11}{10}u$.
(c) Show that $e = \frac{1}{2}$.
At the instant of collision, A and B are at a distance d from a vertical barrier fixed to the surface at right angles to their direction of motion. Given that B hits the barrier, and that the coefficient of restitution between B and the barrier is $\frac{11}{16}$,
(d) find the distance of A from the barrier at the instant that B hits the barrier
(e) show that, after B rebounds from the barrier, it collides with A again at a distance $\frac{5}{32}d$ from the barrier. [E]

20 A light inextensible string of length $4a$ has a particle A, of mass m, attached at one end and a particle B, of mass m, attached at the other end. The string passes through a small smooth ring which is fixed at a point O at a distance $3a$ above a horizontal table. The system is hanging in equilibrium with $OB = 2a$ when a smooth bead of mass $2m$, which is threaded on the string between O and B, is released from rest at O. The bead falls under gravity until it collides with and adheres to the particle B to form a composite particle C.

(a) Given that the string remains taut, show that the speed of C immediately after the collision is $\sqrt{(ga)}$.

(b) Find the speed of C immediately before it reaches the table. [E]

21 A cricket match is played on level ground. Just before the batsman hits the ball, it is 1 m above the ground and travelling with speed 20 m s^{-1} in a direction inclined at $\frac{\pi}{6}$ below the horizontal. Immediately after the ball has been hit it has speed 25 m s^{-1} and the direction of motion is reversed. The ball has mass 0.15 kg.

(a) Find the magnitude of the impulse on the ball due to the impact.

Given that the ball first reaches the horizontal ground at a point A, find

(b) the horizontal distance of A from the batsman,

(c) the speed of the ball when it reaches A.

(d) Write down two assumptions which you have made about the forces acting on the ball during its motion. [E]

22 Two particles, A and B, of masses $2m$ and $3m$ respectively, are moving in a straight line in the same direction on a smooth horizontal plane. The particles collide and, *after* the collision, A and B continue to move in the same straight line and in the same direction with speeds u and $3u/2$ respectively. Given that the coefficient of restitution between A and B is $\frac{1}{5}$, find, in terms of u, the speed of A and the speed of B *before* their collision. Find also, in terms of m and u, the magnitude of the impulse of the force exerted by B on A during the collision. [E]

23 Three identical smooth spheres, A, B and C, each of mass m, lie at rest in a straight line on a smooth horizontal table. Sphere A is projected with speed u to strike sphere B directly. Sphere B then strikes sphere C directly. The coefficient of restitution between any two spheres is e. Find the speeds, in terms of u and e, of the spheres after these two collisions. Show also that the total loss of kinetic energy is

$$\frac{mu^2}{16}[5 + 2e - 4e^2 - 2e^3 - e^4].$$ [E]

Describe a physical situation that could be modelled by the above problem.

24 A smooth horizontal rail is fixed at a height $3h$ above rough horizontal ground. A uniform rod AB, of mass M and length $6h$, is placed in a vertical plane perpendicular to the rail with the end A resting on the ground. The distance $AC = 5h$, where C is the point of contact between the rail and the rod. Show that the force exerted by the rail on the rod is of magnitude $12Mg/25$.
Given that equilibrium is limiting, find the coefficient of friction between the rod and the ground and show that the force exerted by the ground on the rod is of magnitude $17Mg/25$.
Find, in terms of M and g, the greatest magnitude of the horizontal force which could be applied to the rod at A without disturbing equilibrium. [E]

25 A uniform straight rod AB, of mass M and length $2l$, rests in limiting equilibrium with the end A on rough horizontal ground and the end B against a smooth vertical wall. The vertical plane containing AB is at right angles to the wall and the coefficient of friction between the rod and the ground is $\frac{1}{3}$.
(a) Show that AB is inclined at $\arctan\left(\frac{3}{2}\right)$ to the horizontal.
With AB in the same position, a horizontal force of magnitude kMg is applied to the mid-point of the rod towards the wall so that the line of action of this force is at right angles to the wall.
Given that equilibrium is limiting with the end A on the point of moving towards the wall, calculate
(b) the value of k,
(c) the magnitude of the resultant force exerted by the rod on the ground, giving your answer in terms of M and g. [E]

26

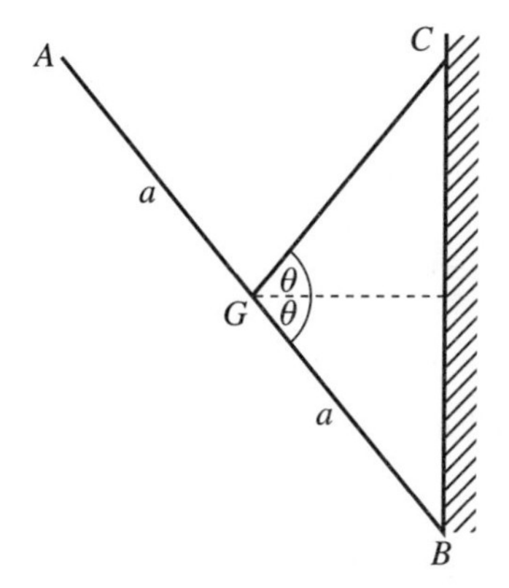

The diagram shows a uniform rod AB of mass M and length $2a$ with the end B in contact with a rough vertical wall. The rod is held in equilibrium by a light inextensible string of length a with one end attached to the mid-point G of the rod and the other end attached to a point C of the wall vertically above B. The string and the rod each make an acute angle θ with the horizontal.

(a) Show that μ, the coefficient of friction between the wall and the rod, satisfies the relation $\mu \geqslant \tan\theta$.

When a force of magnitude $\frac{1}{2}Mg$ acts at A, in the direction from A to C, the system remains in equilibrium.

(b) Find, in terms of M, g and θ, expressions for the horizontal and vertical components of the force acting on the rod at B.

(c) Given that $\mu = 0.75$, deduce that $\tan\theta \leqslant \frac{1}{2}$. [E]

27 A smooth cylinder of radius a is fixed on a rough horizontal table with its axis parallel to the table. A uniform rod ACB of length $6a$ and mass M rests in limiting equilibrium with the end A on the table and the point C touching the cylinder. The vertical plane containing the rod is perpendicular to the axis of the cylinder and the rod makes an angle 2θ with the table.

(a) Show that the magnitude of the force exerted by the cylinder on the rod is

$$3Mg\cos 2\theta \tan\theta.$$

(b) Show also that μ, the coefficient of friction between the rod and the table, is given by

$$\mu(\cot\theta - 3\cos^2 2\theta) = 3\sin 2\theta \cos 2\theta.$$ [E]

28 A uniform ladder of length $4l$ and mass M rests with one end A on rough horizontal ground and the other end B against a smooth vertical wall. The vertical plane containing AB is at right angles to the wall. The coefficient of friction between the ladder and the ground is $\frac{1}{5}$. A particle of mass $2M$ is attached to the ladder at C where $AC = 3l$.
Given that equilibrium is limiting, show that the ladder is inclined at an angle $\arctan\left(\frac{10}{3}\right)$ to the horizontal.
The ladder is moved to a similar position where the wall is rough. The ladder rests at an angle $\arctan\left(\frac{10}{3}\right)$ to the horizontal and the coefficients of friction between the ladder and the wall and the ladder and the ground are both $\mu (\neq \frac{1}{5})$.
The particle of mass $2M$ is moved to the top of the ladder.
Given that equilibrium is limiting, show that

$$\mu^2 + 20\mu - 5 = 0.$$ [E]

Obtain the value of μ to 3 decimal places.
Describe a physical situation that this problem could be used to model.
State the assumptions made in producing this model.
Interpret your final answer in terms of the original situation.

29

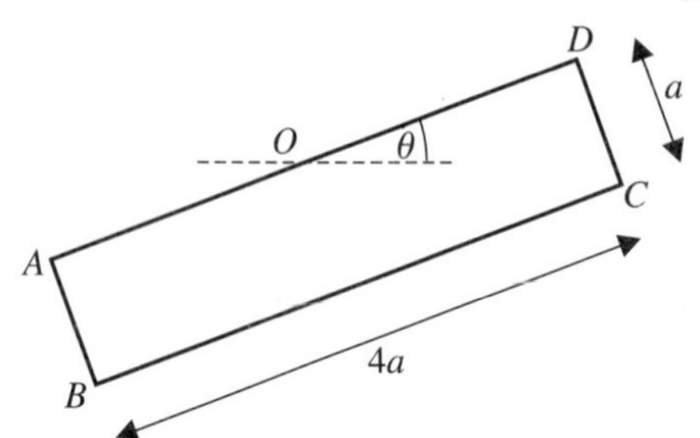

The diagram shows a uniform rectangular lamina $ABCD$, free to rotate in a vertical plane about a smooth horizontal axis through O, the mid-point of AD. The lamina is of mass $2m$ and the sides AB and BC are of length a and $4a$ respectively. The angle θ is the inclination of OD to the horizontal. Particles of mass $5m$, $2m$ and $4m$ are attached at the points A, B and D respectively.
Find, in terms of m, a, g and θ, an expression for the anti-clockwise moment about O of the four weights.
Given that the system is in equilibrium, show that $\tan\theta = 2$.
When a force of magnitude P is applied at the point C and in the direction $\overrightarrow{CB}$ the system rests in equilibrium with AD horizontal and uppermost. Show that $P = 6mg$ and, in this position, find the magnitude of the reaction at O. [E]

30 A uniform rod AB, of mass m and length l, is smoothly jointed at the end A to a fixed straight horizontal wire AC. The end B is attached by means of a light inextensible string, also of length l, to a small ring, of mass m, which can slide on the wire, the coefficient of friction between the ring and the wire being μ. The rod is in equilibrium in the vertical plane through the wire. Given that α is the inclination of the string to the horizontal, show that the tension in the string is of magnitude $mg/(4 \sin \alpha)$.
Show also that

$$\tan \alpha \geqslant 1/(5\mu).$$ [E]

31 A uniform ladder, of mass M and length l, rests in a vertical plane with one end against a smooth vertical wall, the wall being perpendicular to the vertical plane through the ladder. The other end of the ladder rests on horizontal ground, the coefficient of friction between the ladder and the ground being 1/4. The ladder is inclined at an angle θ to the horizontal, where $\tan \theta = 24/7$. A man of mass $10M$ climbs up the ladder. Show that the man can reach a height of $6l/7$ above the ground before the ladder begins to slip. [E]
What assumptions have you made in modelling the above physical situation in order to obtain the mathematical model you solved?
How could this model be refined so that it becomes a better model of the physical situation?

32 A uniform ladder, of mass 45 kg and length 4 m, rests in a vertical plane with one end against a smooth vertical wall, the wall being perpendicular to the vertical plane through the ladder. The other end of the ladder rests on horizontal ground, the coefficient of friction between the ladder and the ground being 0.4. The ladder is inclined at an angle θ to the horizontal where $\tan \theta = 1.5$. A man of mass 60 kg climbs up the ladder When he reaches a point x m from the bottom of the ladder, measured along the ladder, the ladder slips.
Model the above situation and hence obtain the value of x.
State clearly the assumptions you have made in producing your model. How can your model be refined so as to take into account further properties of the actual situation?

33 A uniform metal rod AB of length 3 m and mass 20 kg is smoothly hinged at A to a vertical wall. It is held in a horizontal position by a cord fixed at the end B of the rod. The other end of the cord is fixed to a point of the wall vertically above A so that the cord makes an angle of 30° to the horizontal. A lamp of mass 30 kg is suspended from B.

(a) Model this situation and hence obtain an estimate for the tension in the cord. State your modelling assumptions.

The breaking tension of the cord is 1960 N.

(b) It is required to hang an additional piece of equipment of mass 90 kg from the rod. Find the greatest possible distance from A that this can be hung without the cord breaking.

34 A uniform rod, of mass M, rests in limiting equilibrium with the end A standing on rough horizontal ground and the end B resting against a smooth vertical wall. The vertical plane containing AB is perpendicular to the wall. The coefficient of friction between the rod and the ground is 0.2.

Find, to the nearest degree, the angle at which the rod is inclined to the vertical. [E]

35

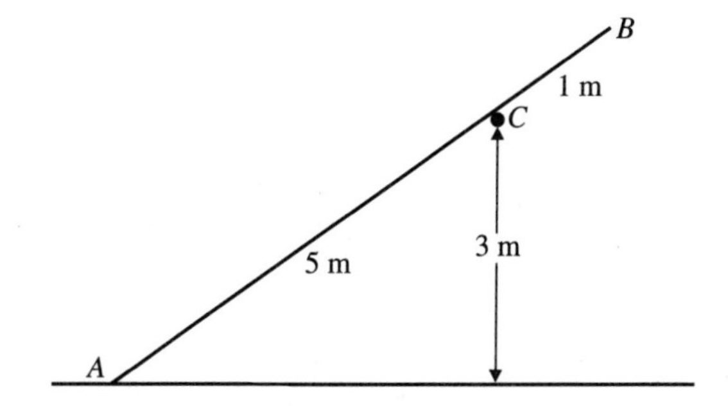

A smooth horizontal rail is fixed at a height of 3 m above a horizontal playground whose surface is rough. A straight uniform pole AB, of mass 20 kg and length 6 m, is placed to rest at a point C on the rail with the end A on the playground. The vertical plane containing the pole is at right angles to the rail. The distance AC is 5 m and the pole rests in limiting equilibrium.

Calculate:

(a) the magnitude of the force exerted by the rail on the pole, giving your answer to the nearest N,

(b) the coefficient of friction between the pole and the playground, giving your answer to 2 decimal places,

(c) the magnitude of the force exerted by the playground on the pole, giving your answer to the nearest N. [E]

36

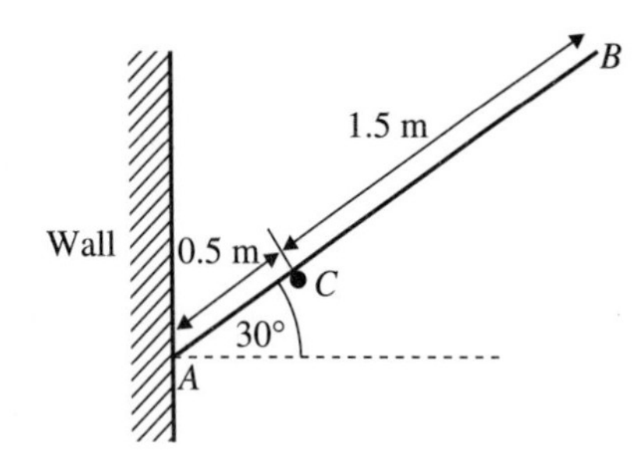

The diagram shows a uniform heavy rod AB of mass 0.5 kg and length 2 m. The rod rests in equilibrium at 30° to the horizontal, with its lower end A in contact with a rough vertical wall. It is supported by a smooth horizontal rail at C, a point of the rod, where $AC = 0.5$ m. The vertical plane containing the rod is perpendicular to the plane of the wall. Find:

(a) the magnitude of the force, in N to 1 decimal place, exerted by the rail on the rod at C,

(b) the magnitudes, in N, of the horizontal and vertical components of the force exerted by the wall on the rod at A.

Given that the coefficient of friction between the rod and the wall is μ,

(c) find the minimum value of μ. [E]

Examination style paper

M2

Answer all questions **Time allowed 90 minutes**

Whenever a numerical value of g is required, take $g = 9.8\ \mathrm{ms}^{-2}$

1 A cricket ball of mass 0.5 kg is moving with velocity $(2\mathbf{i} - 3\mathbf{j})\ \mathrm{m\,s}^{-1}$ when it is struck by a cricket bat. After the impact the ball has velocity $(6\mathbf{i} + 2\mathbf{j})\ \mathrm{m\,s}^{-1}$. Find to 3 significant figures the magnitude of the impulse exerted by the bat on the ball.

(5 marks)

2 A lorry of mass 4500 kg is travelling up a straight road inclined at an angle α to the horizontal where $\sin\alpha = \frac{1}{100}$. The engine of the lorry is working at a steady 40 kW and the non-gravitational resistance to motion is constant and of magnitude 1000 N. At the instant when the lorry is moving with a speed of $15\ \mathrm{m\,s}^{-1}$ calculate, in $\mathrm{m\,s}^{-2}$ to 3 significant figures, the acceleration of the lorry.

(6 marks)

3 A small smooth sphere A of mass 2 kg is travelling along a straight line on a smooth horizontal plane with speed $4\ \mathrm{m\,s}^{-1}$ when it collides with a small smooth sphere B of mass 1 kg moving along the same line in the same direction with speed $2\ \mathrm{m\,s}^{-1}$. After the collision A continues to move in the same direction with speed $3\ \mathrm{m\,s}^{-1}$.

(*a*) Find the speed of B after the collision.

(3 marks)

(*b*) Find the coefficient of restitution between A and B.

(3 marks)

(*c*) State any assumptions you have made in your calculation.

(1 mark)

4

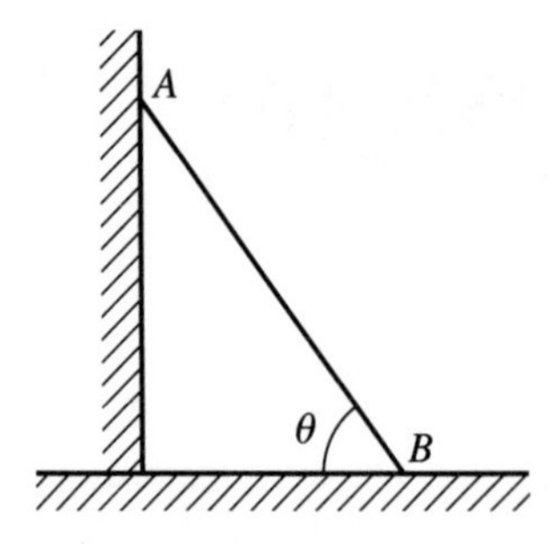

The diagram shows a uniform ladder AB resting in limiting equilibrium with its top A against a smooth vertical wall and its base B on a rough horizontal floor. The coefficient of friction between the ladder and the floor is $\frac{3}{4}$. Given that the ladder makes an angle θ with the floor show that

$$\tan\theta = \tfrac{2}{3}.$$ **(8 marks)**

5 A particle P moves along the x-axis. It passes through the origin O at time $t = 0$ with speed $12\,\text{m}\,\text{s}^{-1}$ in the direction of x increasing. At time t seconds the acceleration of P, in the direction of x increasing, is $(6t - 15)\,\text{m}\,\text{s}^{-2}$.

(*a*) Find the values of t at which P is instantaneously at rest.

(5 marks)

(*b*) Find the distance between the points at which P is instantaneously at rest.

(5 marks)

6

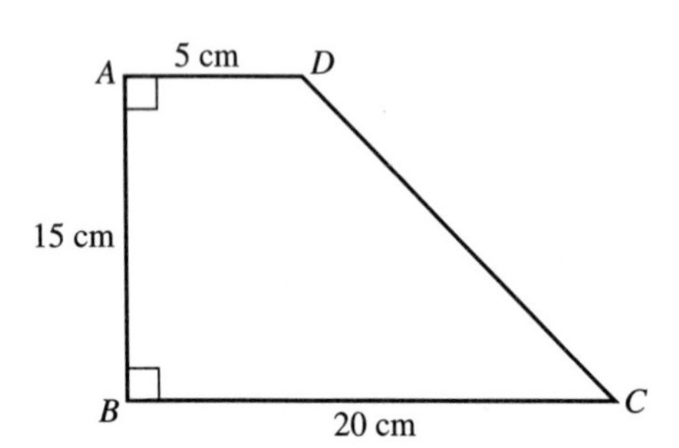

The diagram shows a plane uniform lamina $ABCD$ formed from a uniform rectangular lamina by removing a triangle. The lengths of AB, AD and BC are 15 cm, 5 cm and 20 cm respectively.

(*a*) Find the distance of the centre of mass

(i) from AB,

(ii) from BC.

(9 marks)

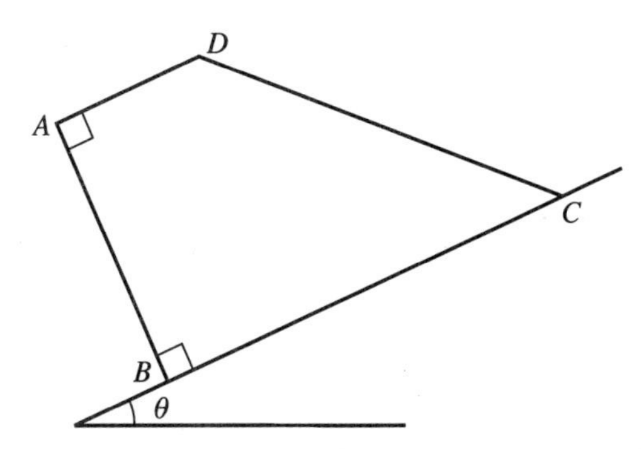

The lamina is placed on a plane inclined at angle θ to the horizontal as shown above.

The plane is sufficiently rough to prevent slipping.

(*b*) Determine the maximum value of θ for the lamina to remain in equilibrium in this position.

(3 marks)

7

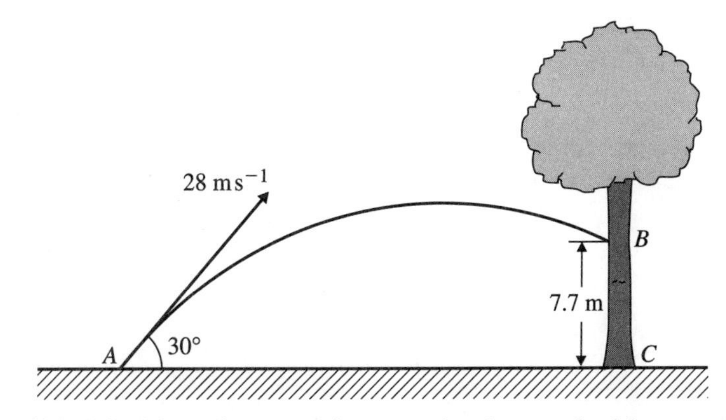

A golf ball is driven from a point A on level ground with a speed of $28\,\text{m}\,\text{s}^{-1}$ at an angle of elevation of 30°. The golf ball is modelled by a particle moving freely under constant gravity.

(*a*) Find the greatest height achieved by the ball above the level of A.

(4 marks)

On its downward flight the ball hits a vertical tree CB at the point B which is 7.7 m above the level of A as shown in the figure.

(*b*) Find the time in seconds to 3 significant figures taken by the ball to reach B from A.

(4 marks)

(*c*) Find the distance AC in metres to 3 significant figures.

(3 marks)

(*d*) Discuss briefly two assumptions you have made in your calculation and give reasons why they are justified in this case.

(2 marks)

8 A particle P of mass 2 kg is acted on by two forces $\mathbf{F}_1 = (4\mathbf{i} + 3\mathbf{j})$ N and $\mathbf{F}_2 = (2\mathbf{i} - \mathbf{j})$ N.

(*a*) Show that the acceleration of P is $(3\mathbf{i} + \mathbf{j})\,\text{m s}^{-2}$.

(3 marks)

At time $t = 0$ P is at the point with position vector $(2\mathbf{i} - \mathbf{j})$ m relative to a fixed origin O and has velocity $4\mathbf{i}\,\text{m s}^{-1}$.

Calculate at time $t = 3$ seconds

(*b*) the velocity of P,

(5 marks)

(*c*) the position vector of P,

(4 marks)

(*d*) the kinetic energy of P.

(2 marks)

Answers

Note on accuracy: Unless otherwise stated all numerical answers are given to 3 significant figures. When calculating answers the maximum number of figures possible should be used before the final answer is rounded to 3 significant figures. Rounding intermediate answers can produce an inaccurate final answer.

Exercise 1A

1 3.50 s, 52.5 m **2** 30.6 m
3 11.7 m s^{-1} **4** 79.9 m
5 2.67s, 80.2 m **6** 14 m s^{-1}, 1.23 m
7 $(6\mathbf{i} + 13.0\mathbf{j})$ m **8** 0, 21.6
9 14 m s^{-1}
10 (a) 4.43 m s^{-1} (b) 14.7 m s^{-1}, 17.6° below horizontal
11 3.68 m
12 25 m s^{-1}, 14.7 m s^{-1}, 29 m s^{-1}, 30.5° below horizontal
13 $(14\mathbf{i} - 17.6\mathbf{j})$ m
14 6 m s^{-1}, 2 s **15** 123 m **16** 9.49 m
17 16.8 m s^{-1}, 4.97 m **18** Yes, 0.475 m

Exercise 1B

1 (a) 3.54 s (b) 61.3 m (c) 7.07 s (d) 245 m
2 (a) 2.86 s (b) 40 m (c) 5.71 s (d) 277 m
3 36.3° **4** (a) 13.4 m s^{-1}, 38.6° above horizontal (b) 10.6 m s^{-1}, 7.67° below horizontal (c) 15.4 m s^{-1}, 46.9° below horizontal.
5 8.03 m
6 7 m, 7.22 m, 18.4° above horizontal
7 $(10\mathbf{i} - 12.4\mathbf{j})$ m, 15.5 m
8 $(8\mathbf{i} + 22.4\mathbf{j})$ m, $(4\mathbf{i} + 1.4\mathbf{j})$ m s^{-1}
9 235 m. Point of projection on same horizontal level as point of impact.
10 7.8 m s^{-1}, 0.776 m
11 8.64 m
12 35 600 m
13 5.45 s, 87.1 m
14 4.82 m, 80.4° above horizontal
15 104 m
16 11.6° above horizontal
17 5.59
18 $(5\mathbf{i} + 12.3\mathbf{j})$ m s^{-1}
19 $(10\mathbf{i} - 4.6\mathbf{j})$ m s^{-1}, 22.5 m
20 4.04 s, 21.8° above horizontal
21 (a) 25 m s^{-1} (b) 1.6 m (c) 1.22 m
22 31.9 m, 3.12 s, 221 m

Exercise 1C

1 (a) 28 m s^{-1} (b) 14 m s^{-2}
2 (a) $t = 0$ s, $\frac{4}{9}$ s (b) 0 m, 0.132 m
3 (a) 0 m s^{-1} (b) -45 m s^{-2} (c) 2 m s^{-1} (d) -8 m
4 (a) 1 m s^{-1} (b) $\frac{2}{3}$ m (c) 1 s (d) $\frac{1}{3}$ m
5 (a) 0 s, $\frac{3}{5}$ s (b) 4 m, 3.82 m (c) 17 m s^{-2}
6 (a) $5\frac{2}{3}$ m (b) 2 m s^{-1} (c) $4\frac{1}{3}$ m
7 3 s
8 (a) $14\frac{2}{3}$ m (b) $68\frac{2}{3}$ m

Exercise 1D

1 (a)

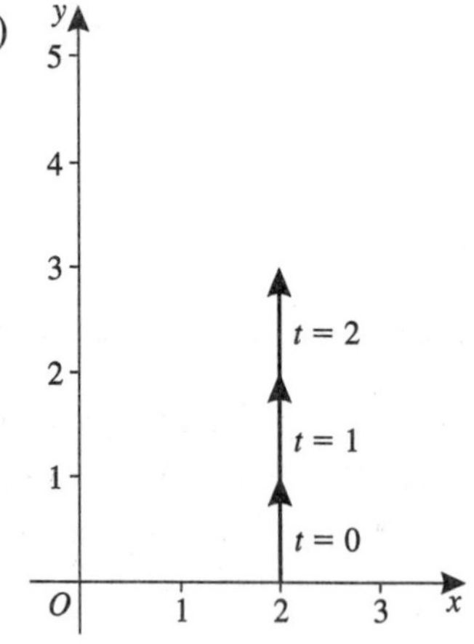

(b)

(c)

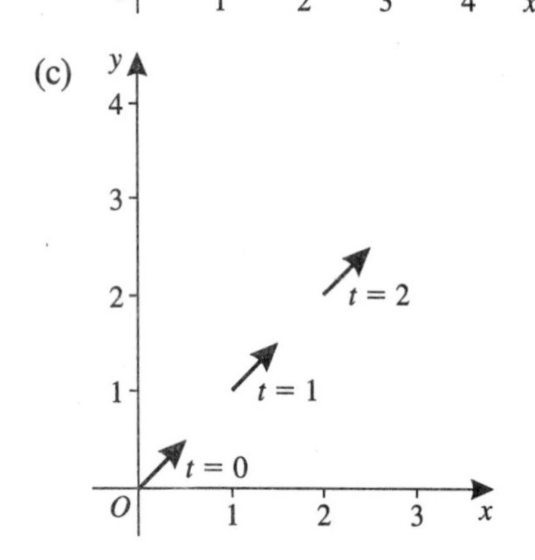

(d)

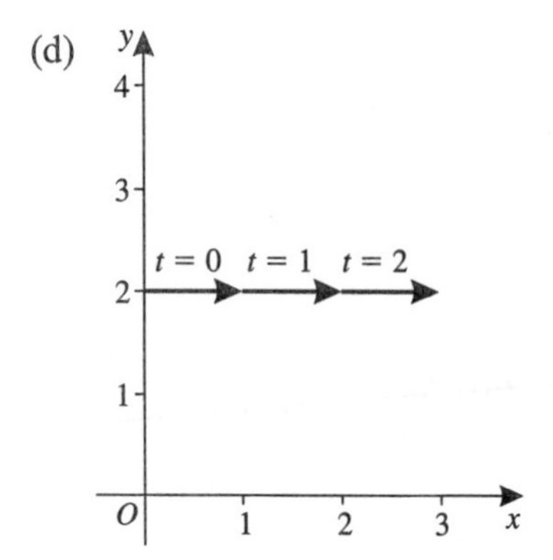

(e)

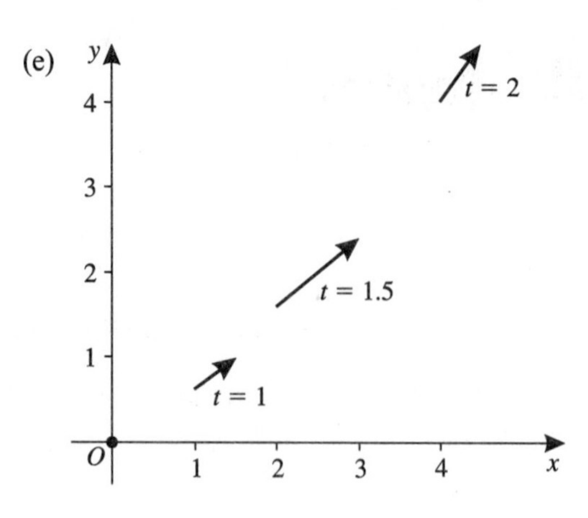

2 (a) $\mathbf{j}$; $\mathbf{j}$, 1 (b) $2t\mathbf{j}$; $2\mathbf{j}$, 2 (c) $\mathbf{i}+\mathbf{j}$; $\mathbf{i}+\mathbf{j}$, $\sqrt{2}$
(d) $\mathbf{i}$; $\mathbf{i}$, 1 (e) $2t\mathbf{i}+\frac{3}{2}t^2\mathbf{j}$; $(2\mathbf{i}+\frac{3}{2}\mathbf{j})$, $\frac{5}{2}$

3 (a) $\mathbf{i}+2t\mathbf{j}$; $2\mathbf{j}$ (b) $t\mathbf{i}+\mathbf{j}$; $\mathbf{i}$ (c) $3t^2\mathbf{i}+\mathbf{j}$, $6t\mathbf{i}$
(d) $3t^2\mathbf{i}$; $6t\mathbf{i}$ (e) $t^2\mathbf{i}+t^3\mathbf{j}$; $2t\mathbf{i}+3t^2\mathbf{j}$

4 (a) $2t\mathbf{i}$, $2\mathbf{i}$ (b) $2t\mathbf{i}+\mathbf{j}$, $2\mathbf{i}$
(c) $2t\mathbf{i}+2t\mathbf{j}$, $2\mathbf{i}+2\mathbf{j}$
(d) $3t^2\mathbf{i}+3t^2\mathbf{j}$, $6t\mathbf{i}+6t\mathbf{j}$
(e) $t^3\mathbf{i}+t^2\mathbf{j}$, $3t^2\mathbf{i}+2t\mathbf{j}$

5 (a) 4 (b) 4.12 (c) 5.66 (d) 17.0 (e) 8.94

6 (a) $t^2\mathbf{i}+2t\mathbf{j}$; $2\mathbf{i}$ (b) $t^2\mathbf{i}+t^3\mathbf{j}$; $2\mathbf{i}+6t\mathbf{j}$
(c) $t\mathbf{i}+t^3\mathbf{j}$; $6t\mathbf{j}$ (d) $t^4\mathbf{i}+t^3\mathbf{j}$; $12t^2\mathbf{i}+6t\mathbf{j}$
(e) $\frac{1}{3}t^3\mathbf{i}+3t^4\mathbf{j}$; $2t\mathbf{i}+36t^2\mathbf{j}$

7 (a) $(t+1)\mathbf{i}$; $(\frac{1}{2}t^2+t)\mathbf{i}$ (b) $\mathbf{i}+t\mathbf{j}$; $t\mathbf{i}+\frac{1}{2}t^2\mathbf{j}$
(c) $(t+1)\mathbf{i}+t\mathbf{j}$; $(\frac{1}{2}t^2+t)\mathbf{i}+\frac{1}{2}t^2\mathbf{j}$
(d) $(3t^2+1)\mathbf{i}+t\mathbf{j}$; $(t^3+t)\mathbf{i}+\frac{1}{2}t^2\mathbf{j}$
(e) $(t+1)\mathbf{i}+3t^2\mathbf{j}$; $(\frac{1}{2}t^2+t)\mathbf{i}+t^3\mathbf{j}$

8 (a)

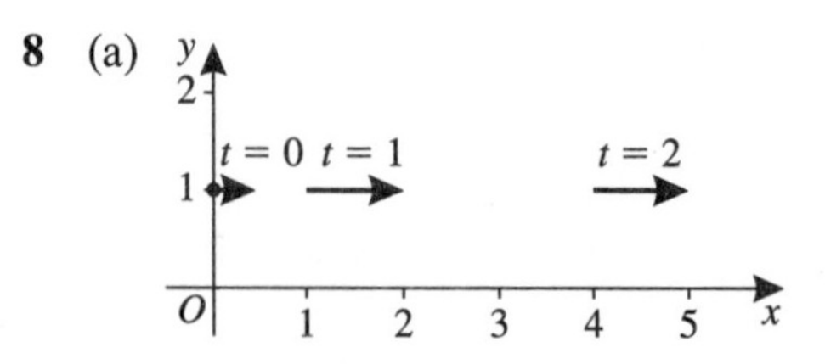

(b)

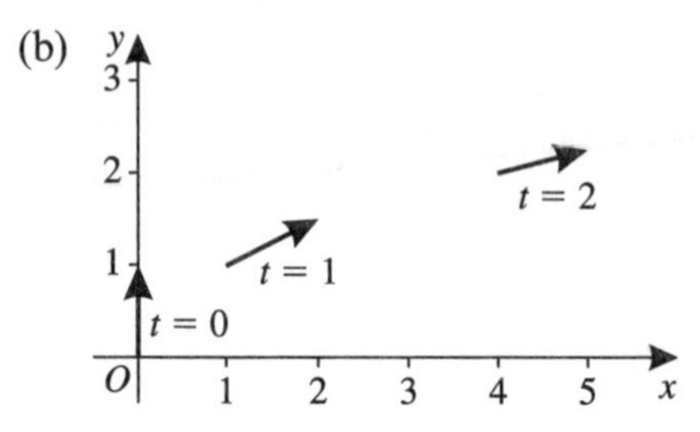

(c)

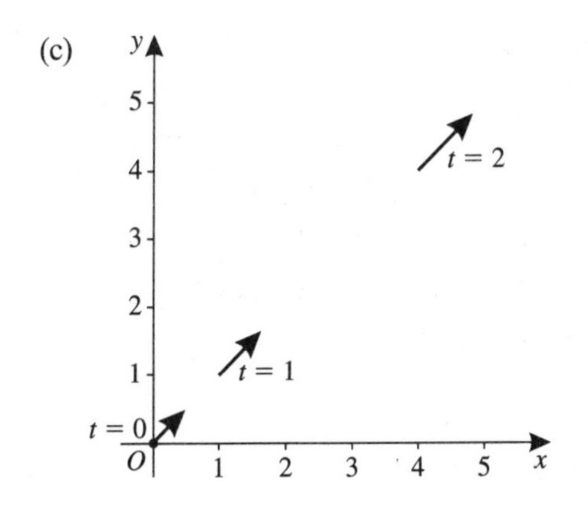

9 $(6t-4)\mathbf{i}+4t\mathbf{j}$

10 (a) $(16\mathbf{i}+6\mathbf{j})\,\text{m s}^{-2}$ (b) $(\frac{2}{3}\mathbf{i}+\frac{4}{5}\mathbf{j})$ m
(c) 1.04 m

11 (a) $(12\mathbf{i}-8\mathbf{j})\,\text{m s}^{-1}$ (b) $14.4\,\text{m s}^{-1}$
(c) $(\mathbf{i}-2\mathbf{j})$ m $(8\mathbf{i}-8\mathbf{j})$ m (d) 9.22 m

12 $18.3\,\text{m s}^{-1}$

13 $5\frac{5}{6}$ m

14 (a) $((2t^3+3)\mathbf{i}+(t^2+3t+4)\mathbf{j})$ m
(b) $((3t+2)\mathbf{i}+(10-2t)\mathbf{j})$ m (c) $(5\mathbf{i}+8\mathbf{j})$ m

15 (a) $((4t-p)\mathbf{i}+(3t^2-3)\mathbf{j})\,\text{m s}^{-1}$ (b) 1 s, 4

16 $((2t+12)\mathbf{i}+4\mathbf{j})\,\text{m s}^{-1}$, $(2t\mathbf{i}+2\mathbf{j})\,\text{m s}^{-1}$, 6 s

17 (a) 20.6 m (b) 10 m s^{-1}, 53.1° below **i**
(c) $\mathbf{i}\,\text{m s}^{-2}$

Exercise 1E

1 (a) $15\,\text{m s}^{-1}$ (b) $\left(\frac{1}{3}t^3-\frac{1}{2}t^2+3t\right)$ m
(c) $10\frac{2}{3}$ m

2 (a) 2.07 s (b) 28.4 m (c) air resistance

3 (a) 84.9 N
(b) $(t^2+3t+3)\mathbf{i}+(2-2t^2)\mathbf{j}\,\text{m s}^{-1}$
(c) 1

4 (a) $24.3\,\text{m s}^{-1}$ (b) 9.46° (c) 9.49 N

5 (a) 3, 4 (b) $5\frac{5}{12}$ m

6 (a) 1.97 m (b) $19.7\,\text{m s}^{-1}$
(c) 23.2° below horizontal

7 (a) $(6t\mathbf{i}+5\mathbf{j})\,\text{m s}^{-2}$ (b) 2 (c) 6.10 m

8 (a) $5\,\text{m s}^{-1}$ (b) $2\frac{1}{7}$ m
(c) $1.96\,\text{m s}^{-2}$ (d) 0.510 s

9 (a) $19.0\,\text{m s}^{-1}$ (b) $t=2$
(c) 40.2 N, $\theta=63°$

10 (a) 54 m (b) 42.9 m

(c) assumptions – golf ball may be treated as a particle, air resistance may be neglected, g may be taken as constant – justified because ball is small, air resistance has a little effect on a small body, only slight variations in g over these heights.

Exercise 2A

1 $(4\frac{1}{6}, 0)$ **2** $(0, 4\frac{2}{7})$ **3** (3,3) **4** $1\frac{2}{3}$ m

5 6 **6** 7 kg **7** (1, 4) **8** (0.5, 0.25)

9 (a) 1.91 m (b) 1.82 m

10 (a) 4 cm (b) $4\frac{4}{9}$ cm

11 (a) 12 cm (b) 2 cm

12 2 kg, $1\frac{1}{7}$ cm **13** (a) 1.13 m (b) 0.467m

Exercise 2B

1 (a) (2.5, 1.5) (b) (4.5, 0.5) (c) (6, 6)
(d) (3, 0) (e) $(13\frac{1}{3}, 10)$ (f) $(8, 9\frac{1}{3})$

2 (a) $(2\frac{3}{7}, 2\frac{1}{14})$ (b) $(1\frac{2}{3}, 2\frac{5}{6})$ (c) $(3\frac{2}{3}, 0)$
(d) (7.12, 6.88) (e) (1.97, 2.03)
(f) (7.05, 6.05) (g) $(3, 2\frac{1}{5})$ (h) (3, 3.94)

3 (a) 3.60 cm (b) 5.75 cm (c) 7.64 cm
(d) 4.77 cm (e) 6.62 cm

4 (a) 3.82 cm (b) 3.82 cm (c) 5.28 cm
(d) 7.20 cm (e) 6.50 cm

5 (a) 0.75 m (b) 0.5 m

6 (a) 10.5 cm (b) 2 cm

7 (a) $1\frac{1}{6}$ m (b) 1 m **8** (2.35, −0.1)

9 On axis of symmetry 6.26 cm from corner of square.

10 5.42 cm

11 2.62 cm **12** 6.62 cm **13** 2.47 cm

14 0.340 m **15** $3\pi : (3\pi-8)$

Exercise 2C

1 58° **2** 14° **3** 49.4°

4 (a) 33.7° (b) 21.8°

5 0.99 cm 26°

6 (a) 6.25 cm
(b) 5 cm, 38.7° Peg is smooth and so no friction. Peg is small.

7 possible, 26.6°

8 (a) yes (b) no 15.8°
9 26.6° sliding. Surface sufficiently rough that there is no sliding for $\theta \leqslant 26.6°$
10 38.7° (a) possible (b) not possible
11 (a) 2.9 cm (b) 3.2 cm (c) 31°
12 0.401 radians
13 42.6°
14 (a) 4.90 (b) 39.2°

Exercise 3A

1 0.84 J 2 5.5 N 3 588 J 4 157 J
5 300 J 6 12.0 m 7 410 J 8 274 J
9 0.25 10 24 000 J 11 196 J 12 567 J
13 153 J, 441 J
14 (a) 118 N (b) 4700 J (c) 14 100 J
15 637 J 16 0.0255

Exercise 3B

1 (a) 10 J (b) 4 J (c) 125 J
(d) 1880 J (e) 108 000 J (f) 813 J
2 (a) 34.3 J gain (b) 12 700 J loss
(c) 235 000 J gain (d) 68 600 J gain
(e) 2160 J gain (f) 265 000 J loss
3 6 J 4 328 000 J 5 8 m s^{-1}
6 3 m s^{-1} 7 945 J 8 80 J
9 216 000 J 10 11.8 J, 10.0 J

Exercise 3C

1 29.4 J, 9.90 m s^{-1}
2 (a) 200 J (b) 200 J (c) 20.4 m
3 64 J (b) 64 J (c) 4 m
4 (a) 75 J (b) 75 J (c) 3.83 m
5 (a) 7.5 J (b) 7.5 J (c) 0.425
6 11.7 m s^{-1} 7 81.6 m
8 20 kN 9 69.4 mm
10 (a) 37.8 J (b) 37.8 J (c) 5.02 m s^{-1}
11 1.63 m 12 12.8 m
13 10.8 m s^{-1} 14 5.31 m
15 27 000 J, 61.2 m 16 19.6 N
17 34.8 m 18 16.3 m s^{-1}
19 4.23 m s^{-1} 20 0.138

Exercise 3D

1 12 kW 2 9.6 kW 3 $266\frac{2}{3}$ N
4 9 m s^{-1} 5 12.6 m s^{-1} 6 500 N
7 (a) $1\frac{1}{3}$ m s^{-2} (b) 0.296 m s^{-2} (c) 20 m s^{-1}
8 400 N 9 9 kW 10 10 m s^{-1}
11 310 N, 0.49 m s^{-2} 12 14 kW, 6.66 m s^{-1}
13 0.2 m s^{-2}, 11.0 m s^{-1}
14 12.0 m s^{-1}, 0.0855 m s^{-2} 15 30.6 m s^{-1}

Exercise 3E

1 (a) 1.11 m s^{-2} (b) 0.253 m s^{-2}
2 (a) 4.53 m s^{-1} (b) 228 J
3 25 400 kW
4 (a) 21 J (b) 21 J (c) 0.84 N
5 (a) 0.697 m s^{-2} (b) 1170 N
6 50.9 kW
7 (a) 3.92 m s^{-2} (b) 1.5 m
8 (a) 20 m s^{-1} (b) 2.35 J
(c) 19.4 m s^{-1} (d) 19.9 m s^{-1}
9 (a) 9.8 N (b) 492 J (c) 12.6 m s^{-1}
10 (a) 40 m s^{-1} (b) 0.16 m s^{-2}

Review exercise 1

1 2 s
2 19 m
3 15 m s^{-1}, 19.6 m s^{-1}
4 (a) 26.1 m (b) 24 m (c) 71°
5 (a) 48 m (b) 116 m (c) 2.5, 20**i** – 5.625**j**
6 (c) 30 m s^{-1}
(d) Modelled as a particle with no air resistance, no wind and gravity assumed constant.
7 (a) 20.2 s (b) 3960 m (c) 279 m s^{-1}
8 (a) (b) 40.3 (c) 62.5 (d) 39.1 ms^{-1}
9 (b) 78°, 10° (c) 2 s
10 (a) 25 m (b) 12.5 m
11 (b) $36\frac{3}{4}$ m
12 (a) $(4 - 2t)$ m s^{-2} (b) 6 s
13 (b) $40\frac{1}{3}$ m s^{-1} (c) 196 m
14 (a) 1.5 (b) (3**i** + 12**j**) m s^{-2}
15 (a) (–10**i** + 8**j**) m s^{-1} (b) 12.8 m s^{-1}
16 (a) 20 m s^{-1} (b) 2.16 N

17 (a) $V = (2t - 4)\mathbf{i} + (3t^2 + 2at)\mathbf{j}\,\text{m s}^{-1}$
(b) a = −3

18 (b) $2.24\,\text{m s}^{-1}$ 27° with positive x-axis
(c) 4 13.6 m

19 (a) $4\,\text{m s}^{-1}(4t + 8)\text{m}$
$(6 - 2t)\,\text{m s}^{-1}(6t - t^2)\,\text{m}$
(b) 7 m (c) $t = 5$

20 3 **21** 1.5a, a

22 (a) 6a (b) $\dfrac{5a}{6}$

23 (a) 3.5 cm (b) $1\frac{2}{3}$ cm **24** $(3\mathbf{i} + 2.5\mathbf{j})$ m

25 (i) (a) $\frac{33}{13}$ cm, $\frac{58}{13}$ cm (b) 84.1°
(ii) Flat body whose thickness is negligible. Only forces are a vertical force at F and weight of lamina.

26 $9\frac{1}{3}$ cm **27** 14° **28** 50°

29 (a) 0.6d (b) d (c) 59°

30 (a) 2 cm (b) $1\frac{2}{3}$ cm (c) 40°

31 (i) (a) $4\frac{2}{7}$ s (b) 60 m (c) $(30\mathbf{i} + 22.5\mathbf{j})$m
(ii) Any two of: there is no air resistance, gravitational force is constant, there is no wind.

32 $10.5\,\text{m s}^{-1}$ **33** 4650 J

34 (a) $\frac{200}{3}$ m
(b) car modelled as particle, resistances negligible.

35 (b) 60° (c) 1.24 s (d) $9.1\,\text{m s}^{-1}$

36 (a) 29.6 J (b) $(4\mathbf{i} + 6\mathbf{j})\,\text{m s}^{-2}$
(c) 2.9 N 34° (d) 16 m

37 (a) 11 m (b) 162 W **38** 12 kW

39 (b) 20 s (c) 912 W (d) $0.5\,\text{m s}^{-2}$

40 (a) 62 kW (b) $\frac{3}{11}\,\text{m s}^{-2}$ (c) $12\,\text{m s}^{-1}$

41 (a) $0.693\,\text{m s}^{-2}$ (b) 7.43 kN (c) 27.7 kN
(d) 277 kW

42 400 N **43** 21, 14

44 (a) 440 kJ (b) $0.25\,\text{m s}^{-2}$ (c) $16\,\text{m s}^{-1}$

Exercise 4A

1 $(12\mathbf{i} - 2\mathbf{j})\,\text{m s}^{-1}$, ice hockey

2 $(-\mathbf{i} - 2\mathbf{j})$ N s **3** $(\mathbf{i} + 9\mathbf{j})\,\text{m s}^{-1}$

4 $I = 18$ **5** $I = 36$

6 24 N s, a lump of putty **7** 52.7 N s

8 $\frac{3}{7}\,\text{m s}^{-1}$, original direction of A. An engine and a carriage of a train set which couple on colliding.

9 $3\frac{1}{2}\mathbf{i}\,\text{m s}^{-1}$

10 $5\,\text{m s}^{-1}$, direction reversed

Exercise 4B

1 (a) $e = \frac{1}{2}$ (b) $e = \frac{1}{4}$ (c) $e = \frac{2}{3}$

2 (a) $v_1 = 0$, $v_2 = 4$ (b) $v_1 = 3$, $v_2 = 4$
(c) $v_1 = 2\frac{1}{2}$, $v_2 = 5$
(d) $v_1 = -2$, $v_2 = 1$ (e) $v_1 = -1$, $v_2 = 0$

3 (a) $4\,\text{m s}^{-1}$ (b) $e = \frac{1}{4}$
(c) Spheres of same size and can be modelled as particles.

4 (a) $\frac{3}{5}$ (b) 7.5 J

5 $e > \frac{1}{4}$

6 Two snooker balls on a snooker table. Assume balls may be treated as particles and that the table is perfectly smooth.

Exercise 4C

1 (a) $e = \frac{1}{2}$ (b) $e = \frac{3}{4}$

2 (a) $v_1 = 4$ (b) $v_2 = 2.5$

3 (a) $u_1 = 10$ (b) $u_2 = 8$

4 (a) $e = \frac{3}{4}$ (b) 12.6 J
(c) Snooker ball striking a cushion at right angles. Puck on ice rink striking boundary normally.

5 $e = \dfrac{1}{\sqrt{2}}$ **7** $e = \frac{1}{2}$ **8** $h = 128$

9 $e = \frac{3}{5}$. Small rubber ball dropped on to smooth pavement or ice rink.

Exercise 4D

1 (a) $x = 2$, $y = 6$, $v = 1\frac{1}{2}$, $w = 4\frac{1}{2}$
(b) $x = \frac{1}{2}$, $y = 2$, $v = \frac{3}{4}$, $w = 1\frac{1}{2}$

2 (a) $6\,\text{m s}^{-1}$

4 (a) $1\,\text{m s}^{-1}$, $2\,\text{m s}^{-1}$ and $8\,\text{m s}^{-1}$ respectively.
(b) No, since (speed C) > (speed B) > (speed A)

5 Three snooker balls on a smooth horizontal table.

6 (b) $\frac{2}{9}$

Exercise 4E

1 (a) $4\,\text{m s}^{-1}$ (b) 2.86 m

2 (a) A $-11.9\,\text{m s}^{-1}$ up
B $6.7\,\text{m s}^{-1}$ down
(b) 41.9%

3 (a) $4.2\,\text{m s}^{-1}$ (b) 72.1 kN

4 (a) $112.5\,\text{m s}^{-1}$ opposite to original
(b) 2320 kJ

5 (a) $6.26\,\text{m s}^{-1}$ (b) $3.13\,\text{m s}^{-1}$ (c) 1.57 N s

6 (a) $\frac{5u}{8}$ (b) $\frac{15mu}{8}$ (c) $\frac{15}{16}mu^2$

7 (a) 2 s (b) 2.4 m (c) $0.3\,\text{m s}^{-1}$
(d) 0.048 J (e) 0.06 N s

8 (b) 0.8

Exercise 5A

1 (a) 139 N (b) 18.9

2 (a) 36.1 N, 32.0 N (b) 2.04 m from A

3 (a) $\tan\theta = \frac{1}{5}$
(b) 1.22 Mg, 35° to the horizontal

4 (a) 14.1 N (b) 7.07 N, 36.8 N

5 (a) 19.6 N
(b) 62.0 N, 71.6° to the horizontal

6 (a) $P = 17.0$ N (b) $R = 25.9$ N

7 (a) $\frac{10g}{3\pi} = 10.4$ N
(b) $\frac{10g}{3\pi} = 10.4$ N, $5g = 49$ N

Exercise 5B

2 84.9 N 3 45°

4 (a) 1176 N, 467 N (b) 0.505

5 $\frac{2}{3}$ 6 (a) 36.9° (b) $\frac{1}{2}$

7 11.4 N

8 (a) 45°
(b) A uniform ladder against a fairly smooth wall
(c) Take into account roughness of wall; model ladder by non-uniform rod.

10 (a) 84.9 N (c) $m = 7\frac{1}{2}$ kg

Review exercise 2

1 (a) $\frac{1}{5}U$ (b) $\frac{2}{5}MU^2$

2 (a) $(k+1)u$ (b) $k = \frac{1}{6}$ or $\frac{1}{2}$

3 (b) $3mu$ Ns

4 (a) $v_A = \frac{u}{4}(3-e)$, $v_B = \frac{3u}{4}(1+e)$ (d) $\lambda = \frac{3}{2}$

6 (a) $2\,\text{m s}^{-1}$, $4\,\text{m s}^{-1}$ (b) 0.6 N s

8 (a) $J = 3mu$ (c) $3mu^2$
(d) $v_B = \frac{u(9-\lambda)}{3(3+\lambda)}$, $v_c = \frac{4u}{\lambda+3}$ (f) $\frac{8mu}{3}$

9 (a) $10u$, $3u$ (b) $12\,mu$

10 $\frac{1}{3}V(1+e)$, $\frac{1}{3}V(1-2e)$, $e = \frac{1}{2}$

11 (b) $e = \frac{1}{2}$ (c) $e = \frac{2}{5}$ or $\frac{3}{5}$

12 (a) $e = \frac{23}{25}$

13 (c) $\frac{2mu}{3}(1+e)$

14 (a) $\frac{u}{3}(7-2e)$, $\frac{u}{3}(7+e)$ (c) $1.2\,mu$

15 (a) $u(1-4k)$

16 (a) $\frac{u}{6}(5e-1)$ (b) $\frac{1}{5} < e < 1$

17 (c) $\frac{1}{11}(4-150e)$ (d) $\frac{2}{75}$

18 (b) $\frac{u}{4}(3e-1)$ (d) $\frac{27mu}{16}$

19 (b) $\frac{u}{5}(7-3e)$ (d) $\frac{5d}{16}$

20 (b) $\sqrt{(2ga)}$

21 (a) 6.75 N s (b) 56.9 m (c) $25.4\,\text{m s}^{-1}$
(d) No air resistance, no wind, gravitational force constant.

22 $u_A = \frac{14u}{5}$, $u_B = \frac{3u}{10}$, $I = \frac{18mu}{5}$

23 $\frac{1}{2}u(1-e)$, $\frac{1}{4}u(1-e^2)$, $\frac{1}{4}u(1+e)^2$
Three snooker balls in a straight line on a horizontal snooker table with a smooth cloth surface.

24 $\frac{72Mg}{125}$

25 (b) $k = \frac{4}{3}$ (c) $\frac{1}{3}\sqrt{10}\,Mg$

26 (b) $\frac{1}{2}Mg\cot\theta$, $\frac{1}{2}Mg(\tan\theta + 1)$

28 $\mu = 0.247$
A man whose weight is twice that of the ladder climbing the ladder safely to the top. Ladder

modelled by uniform rod. Man modelled by a particle. Wall and floor equally rough. Man can climb to top of ladder if μ is greater than 0.247.

29 $mga(6\cos\theta - 3\sin\theta)$, $\sqrt{205}mg$

31 Ladder may be modelled by a uniform rod. Man may be modelled by a particle. No frictional force at the wall. Model could be improved by taking into account the frictional force at the wall since no surface is perfectly smooth.

32 $x = 2.7$
Model ladder as a uniform rod. Model man as a particle. Assume no frictional force at wall. Model can be refined by taking into account the frictional forces at the wall.

33 (a) 784 N (b) $x = 2$ m
Cord light and inextensible. Lamp modelled as particle. Metal rod modelled as a uniform rod.

34 22°

35 (a) 94 N (b) 0.47 (c) 133 N

36 (a) 8.5 N (b) 4.24 N, 2.45 N (c) 0.58

Examination style paper M2

1 3.20 kg m s^{-1}

2 0.272 m s^{-2}

3 (a) 4 m s^{-1} (b) $e = \frac{1}{2}$
(c) Spheres can be modelled as particles.

5 (a) $t = 1$ and $t = 4$ (b) $13\frac{1}{2}$ m

6 (a) (i) 7 cm (ii) 6 cm (b) 49.4°

7 (a) 10 m (b) 2.11 s (c) 51.3 m or 51.2 m
(d) Golf ball is a particle – good as size is small compared to distances involved in problem.
No air resistance – good as golf ball is small.
Gravity is constant – reasonable as g varies only slightly over these heights.

8 (b) $(13\mathbf{i} + 3\mathbf{j})$ m s^{-1}
(c) $(27\frac{1}{2}\mathbf{i} + 3\frac{1}{2}\mathbf{j})$ m
(d) 178 J

List of symbols and notation

The following notation will be used in all Edexcel examinations.

$\in$	is an element of
$\notin$	is not an element of
$\{x_1, x_2, \ldots\}$	the set with elements $x_1, x_2, \ldots$
$\{x : \ldots\}$	the set of all x such that …
$\mathrm{n}(A)$	the number of elements in set A
$\emptyset$	the empty set
e	the universal set
A'	the complement of the set A
$\mathbb{N}$	the set of natural numbers, $\{1, 2, 3, \ldots\}$
$\mathbb{Z}$	the set of integers, $\{0, \pm 1, \pm 2, \pm 3, \ldots\}$
$\mathbb{Z}^+$	the set of positive integers, $\{1, 2, 3, \ldots\}$
$\mathbb{Z}_n$	the set of integers modulo n, $\{0, 1, 2, \ldots, n-1\}$
$\mathbb{Q}$	the set of rational numbers $\left\{\frac{p}{q} : p \in \mathbb{Z}, q \in \mathbb{Z}^+\right\}$
$\mathbb{Q}^+$	the set of positive rational numbers, $\{x \in \mathbb{Q} : x > 0\}$
$\mathbb{Q}_0^+$	the set of positive rational numbers and zero, $\{x \in \mathbb{Q} : x \geqslant 0\}$
$\mathbb{R}$	the set of real numbers
$\mathbb{R}^+$	the set of positive real numbers, $\{x \in \mathbb{R} : x > 0\}$
$\mathbb{R}_0^+$	the set of positive real numbers and zero, $\{x \in \mathbb{R} : x \geqslant 0\}$
$\mathbb{C}$	the set of complex numbers
(x, y)	the ordered pair x, y
$A \times B$	the cartesian product of sets A and B, $A \times B = \{(a, b) : a \in A, b \in B\}$
$\subseteq$	is a subset of
$\subset$	is a proper subset of
$\cup$	union
$\cap$	intersection
$[a, b]$	the closed interval, $\{x \in \mathbb{R} : a \leqslant x \leqslant b\}$
$[a, b)$, $[a, b[$	the interval $\{x \in \mathbb{R} : a \leqslant x < b\}$
$(a, b]$, $]a, b]$	the interval $\{x \in \mathbb{R} : a < x \leqslant b\}$
(a, b), $]a, b[$	the open interval $\{x \in \mathbb{R} : a < x < b\}$
$y R x$	y is related to x by the relation R
$y \sim x$	y is equivalent to x, in the context of some equivalence relation
$=$	is equal to
$\neq$	is not equal to
$\equiv$	is identical to *or* is congruent to

$\approx$	is approximately equal to
$\cong$	is isomorphic to
$\propto$	is proportinal to
$<$	is less than
$\leqslant, \ngtr$	is less than or equal to, is not greater than
$>$	is greater than
$\geqslant, \nless$	is greater than or equal to, is not less than
∞	infinity
$p \wedge q$	p and q
$p \vee q$	p or q (or both)
$\sim p$	not p
$p \Rightarrow q$	p implies q (if p then q)
$p \Leftarrow q$	p is implied by q (if q then p)
$p \Leftrightarrow q$	p implies and is implied by q (p is equivalent to q)
$\exists$	there exists
$\forall$	for all
$a+b$	a plus b
$a-b$	a minus b
$a \times b$, ab, $a.b$	a multiplied by b
$a \div b$, $\frac{a}{b}$, a/b	a divided by b
$\sum_{i=1}^{n} a_i$	$a_1 + a_2 + \ldots + a_n$
$\prod_{i=1}^{n} a_i$	$a_1 \times a_2 \times \ldots \times a_n$
$\sqrt{a}$	the positive square root of a
$\|a\|$	the modulus of a
$n!$	n factorial
$\binom{n}{r}$	the binomial coefficient $\frac{n!}{r!(n-r)!}$ for $n \in \mathbb{Z}^+$; $\frac{n(n-1)\ldots(n-r+1)}{r!}$ for $n \in \mathbb{Q}$
$\mathrm{f}(x)$	the value of the function f at x
$\mathrm{f}: A \to B$	f is a function under which each element of set A has an image in set B
$\mathrm{f}: x \mapsto y$	the function f maps the element x to the element y
f^{-1}	the inverse function of the function f
$\mathrm{g} \circ \mathrm{f}$, gf	the composite function of f and g which is defined by $(\mathrm{g} \circ \mathrm{f})(x)$ or $\mathrm{gf}(x) = \mathrm{g}(\mathrm{f}(x))$
$\lim_{x \to a} \mathrm{f}(x)$	the limit of $\mathrm{f}(x)$ as x tends to a
Δx, δx	an increment of x
$\frac{\mathrm{d}y}{\mathrm{d}x}$	the derivative of y with respect to x
$\frac{\mathrm{d}^n y}{\mathrm{d}x^n}$	the nth derivative of y with respect to x

$f'(x), f''(x), \ldots f^{(n)}(x)$	the first, second, ... nth derivatives of $f(x)$ with respect to x
$\int y \, dx$	the indefinite integral of y with respect to x
$\int_a^b y \, dx$	the definite integral of y with respect to x betweent he limits $x = a$ and $x = b$
$\frac{\partial V}{\partial x}$	the partial derivative of V with respect to x
$\dot{x}, \ddot{x}, \ldots$	the first, second, . . . derivatives of x with respect to t
e	base of natural logarithms
e^x, $\exp x$	exponential function of x
$\log_a x$	logarithm to the base a of x
$\ln x$, $\log_e x$	natural logarithm of x
$\lg x$, $\log_{10} x$	logarithm of x to base 10
sin , cos , tan cosec, sec, cot	the circular functions
arcsin, arccos, arctan arccosec, arcsec, arccot	the inverse circular functions
sinh, cosh, tanh cosech, sech, coth	the hyperbolic functions
arsinh, arcosh, artanh, arcosech, arsech, arcoth	the inverse hyperbolic functions
i, j	square root of -1
z	a complex number, $z = x + iy$
Re z	the real part of z, Re $z = x$
Im z	the imaginary part of z, Im $z = y$
$\|z\|$	the modulus of z, $\|z\| = \sqrt{(x^2 + y^2)}$
arg z	the argument of z, arg $z = \arctan \frac{y}{x}$
z^*	the complex conjugate of z, $x - iy$
$\mathbf{M}$	a matrix $\mathbf{M}$
$\mathbf{M}^{-1}$	the inverse of the matrix $\mathbf{M}$
$\mathbf{M}^{\mathrm{T}}$	the transpose of the matrix $\mathbf{M}$
det $\mathbf{M}$, $\|\mathbf{M}\|$	the determinant of the square matrix $\mathbf{M}$
$\mathbf{a}$	the vector $\mathbf{a}$
$\overrightarrow{AB}$	the vector represented in magnitude and direction by the directed line segment AB
$\hat{\mathbf{a}}$	a unit vector in the direction of $\mathbf{a}$
$\mathbf{i}, \mathbf{j}, \mathbf{k}$	unit vectors in the directions of the cartesian coordinate axes
$\|\mathbf{a}\|$, a	the magnitude of $\mathbf{a}$
$\|\overrightarrow{AB}\|$, AB	the magnitude of $\overrightarrow{AB}$
$\mathbf{a} . \mathbf{b}$	the scalar product of $\mathbf{a}$ and $\mathbf{b}$
$\mathbf{a} \times \mathbf{b}$	the vector product of $\mathbf{a}$ and $\mathbf{b}$

A, B, C, etc	events
$A \cup B$	union of the events A and B
$A \cap B$	intersection of the events A and B
$\mathrm{P}(A)$	probability of the event A
A'	complement of the event A
$\mathrm{P}(A\|B)$	probability of the event A conditional on the event B
$X, Y, R,$ etc.	random variables
$x, y, r,$ etc.	values of the random variables X, Y, R, etc
$x_1, x_2 \ldots$	observations
$f_1, f_2, \ldots$	frequencies with which the observations $x_1, x_2, \ldots$ occur
$\mathrm{p}(x)$	probability function $\mathrm{P}(X = x)$ of the discrete random variable X
$p_1, p_2, \ldots$	probabilities of the values $x_1, x_2, \ldots$ of the discrete random variable X
$\mathrm{f}(x), \mathrm{g}(x), \ldots$	the value of the probability density function of a continuous random variable X
$\mathrm{F}(x), \mathrm{G}(x), \ldots$	the value of the (cumulative) distribution function $\mathrm{P}(X \leqslant x)$ of a continuous random variable X
$\mathrm{E}(X)$	expectation of the random variable X
$\mathrm{E}[\mathrm{g}(X)]$	expectation of $\mathrm{g}(X)$
$\mathrm{Var}(X)$	variance of the random variable X
$\mathrm{G}(t)$	probability generating function for a random variable which takes the values 0, 1, 2, ...
$\mathrm{B}(n, p)$	binomial distribution with parameters n and p
$\mathrm{N}(\mu, \sigma^2)$	normal distribution with mean μ and variance σ^2
μ	population mean
σ^2	population variance
σ	population standard deviation
$\bar{x}$, m	sample mean
s^2, $\hat{\sigma}^2$	unbiased estimate of population variance from a sample, $s^2 = \frac{1}{n-1}\sum(x_i - \bar{x})^2$
ϕ	probability density function of the standardised normal variable with distribution N(0, 1)
Φ	corresponding cumulative distribution function
ρ	product-moment correlation coefficient for a population
r	product-moment correlation coefficient for a sample
$\mathrm{Cov}\ (X, Y)$	covariance of X and Y

Index